SpringerBriefs in Statistics

JSS Research Series in Statistics

The current research of statistics in Japan has expanded in several directions in line with recent trends in academic activities in the area of statistics and statistical sciences over the globe. The core of these research activities in statistics in Japan has been the Japan Statistical Society (JSS). This society, the oldest and largest academic organization for statistics in Japan, was founded in 1931 by a handful of pioneer statisticians and economists and now has a history of about 80 years. Many distinguished scholars have been members, including the influential statistician Hirotugu Akaike, who was a past president of JSS, and the notable mathematician Kiyosi Itô, who was an earlier member of the Institute of Statistical Mathematics (ISM), which has been a closely related organization since the establishment of ISM. The society has two academic journals: the Journal of the Japan Statistical Society (English Series) and the Journal of the Japan Statistical Society (Japanese Series). The membership of JSS consists of researchers, teachers, and professional statisticians in many different fields including mathematics, statistics, engineering, medical sciences, government statistics, economics, business, psychology, education, and many other natural, biological, and social sciences. The JSS Series of Statistics aims to publish recent results of current research activities in the areas of statistics and statistical sciences in Japan that otherwise would not be available in English; they are complementary to the two JSS academic journals, both English and Japanese. Because the scope of a research paper in academic journals inevitably has become narrowly focused and condensed in recent years, this series is intended to fill the gap between academic research activities and the form of a single academic paper. The series will be of great interest to a wide audience of researchers, teachers, professional statisticians, and graduate students in many countries who are interested in statistics and statistical sciences, in statistical theory, and in various areas of statistical applications.

Rizky Reza Fauzi · Yoshihiko Maesono

Statistical Inference Based on Kernel Distribution Function Estimators

 Springer

Rizky Reza Fauzi
Department of Mathematics
Faculty of Information Technology
and Sciences
Parahyangan Catholic University
Bandung, West Java, Indonesia

Yoshihiko Maesono
Department of Mathematics
Faculty of Science and Engineering
Chuo University
Bunkyo, Tokyo, Japan

ISSN 2191-544X   ISSN 2191-5458 (electronic)
SpringerBriefs in Statistics
ISSN 2364-0057   ISSN 2364-0065 (electronic)
JSS Research Series in Statistics
ISBN 978-981-99-1861-4   ISBN 978-981-99-1862-1 (eBook)
https://doi.org/10.1007/978-981-99-1862-1

This Springer imprint is published by the registered company Springer Nature Singapore Pte Ltd.
The registered company address is: 152 Beach Road, #21-01/04 Gateway East, Singapore 189721,
Singapore

Preface

In statistical inference, we usually assume some underlying distributions which include unknown parameter, but known functional form. On the other hand, without assuming specific distribution, rank statistics give us accurate confidence intervals or significance probabilities. For the inference based on the rank statistic, there are many theoretical findings such as consistency, asymptotic normality, etc. For the finite sample, the test statistics are distribution-free, that is the distribution of the statistics does not depend on the underlying distribution. Unfortunately, the distribution of the rank statistics is discrete, and so the obtained results of the inference are not smooth. The typical example is the empirical distribution function.

In order to conquer this problem, the kernel density function estimator is studied, including its theoretical properties. Since distribution function can be calculated from density function through integration, kernel distribution function estimator is defined in the same way. Furthermore, based on kernel estimators of marginal and joint densities, nonparametric kernel regression is constructed. Replacing the empirical distribution by the kernel estimators, many new statistics are proposed and many papers discuss asymptotic properties of them.

In this book, some topics related to kernel methods will be discussed. We study new estimators of distribution functions, quantiles, and mean residual life functions. Smoothed version of some goodness-of-fit tests will also be discussed. We propose new estimators which solve boundary bias problem. The well-known Kolmogorov-Smirnov and Cramer von Mises statistics are vulnerable to the small change and so the boundary problem is important to solve. Based on the bijective transformation, we propose new estimators which solve the boundary problem and show that the powers of new test statistics are higher than ordinal tests. Further, we will discuss the mean squared error of the kernel-type estimators and obtain the Edgeworth expansion for the quantile estimator.

Bandung, Indonesia
Tokyo, Japan

Rizky Reza Fauzi
Yoshihiko Maesono

Contents

Chapter 1
Kernel Density Function Estimator

Abstract Various kernel-type density estimators are discussed here, such as the standard kernel density and gamma kernel density estimators, including the problem that arises when the data is on nonnegative line. Next, a type of gamma density is used as a kernel function and is modified with expansions of exponential and logarithmic functions. The modified gamma kernel density estimator is not only free of the boundary bias but also variance is also in smaller orders. Simulation results that demonstrate the proposed method's performances are also presented.

Nonparametric methods are gradually becoming popular in statistical analysis for analyzing problems in many fields, such as economics, biology, and actuarial science. In most cases, this is because of a lack of information on the variables being analyzed. Smoothing concerning functions, such as density or cumulative distribution, plays a special role in nonparametric analysis. Knowledge on a density function, or its estimate, allows one to characterize the data more completely. Other characteristics of a random variable can be derived from an estimate of its density function, such as the probability itself, hazard rate, mean, and variance value.

Let $X_1, X_2, \ldots, X_n$ be independently and identically distributed random variables with an absolutely continuous distribution function F_X and a density f_X. The kernel density estimator [1, 2] (the standard one) is a smooth and continuous estimator for density functions. It is defined as

$$\widehat{f_X}(x) = \frac{1}{nh} \sum_{i=1}^{n} K\left(\frac{x - X_i}{h}\right), \quad x \in \mathbb{R}, \tag{1.1}$$

where K is a function called a "kernel", and $h > 0$ is the bandwidth, which is a parameter that controls the smoothness of $\widehat{f_X}$. It is usually assumed that K is a symmetric (about 0) continuous nonnegative function with $\int_{-\infty}^{\infty} K(v)dv = 1$, as well as $h \to 0$ and $nh \to \infty$ when $n \to \infty$. It is easy to prove that the standard kernel density estimator is continuous and satisfies all the properties of a density function.

A typical general measure of the accuracy of $\widehat{f_X}(x)$ is the mean integrated squared error, defined as

© The Author(s), under exclusive license to Springer Nature Singapore Pte Ltd. 2023
Rizky Reza Fauzi and Y. Maesono, *Statistical Inference Based on Kernel Distribution Function Estimators*, JSS Research Series in Statistics,
https://doi.org/10.1007/978-981-99-1862-1_1

$$MISE(\widehat{f}_X) = E\left[\int_{-\infty}^{\infty} \{\widehat{f}_X(x) - f_X(x)\}^2 dx\right].$$

For point-wise measures of accuracy, bias, variance, and the mean squared error $MSE[\widehat{f}_X(x)] = E[\{\widehat{f}_X(x) - f_X(x)\}^2]$ will be used. It is well known that the MISE and the MSE can be computed with

$$MISE(\widehat{f}_X) = \int_{-\infty}^{\infty} MSE[\widehat{f}_X(x)]dx,$$

$$MSE[\widehat{f}_X(x)] = Bias^2[\widehat{f}_X(x)] + Var[\widehat{f}_X(x)].$$

Under the condition that f_X has a continuous second order derivative f_X'', it has been proved by the above-mentioned authors that, as $n \to \infty$,

$$Bias[\widehat{f}_X(x)] = \frac{h^2}{2} f_X''(x) \int u^2 K(u) du + o(h^2), \tag{1.2}$$

$$Var[\widehat{f}_X(x)] = \frac{f_X(x)}{nh} \int K^2(u) du + o\left(\frac{1}{nh}\right). \tag{1.3}$$

There have been many proposals in the literature for improving the bias property of the standard kernel density estimator. Typically, under sufficient smoothness conditions placed on the underlying density f_X, the bias is reduced from $O(h^2)$ to $O(h^4)$, and the variance remains in the order of $n^{-1}h^{-1}$. Those methods that could potentially have a greater impact include bias reduction by geometric extrapolation [3], variable bandwidth kernel estimators [4], variable location estimators [5], nonparametric transformation estimators [6], and multiplicative bias correction estimators [7]. One also could use, of course, the so-called higher order kernel functions, but this method has a disadvantage in that negative values might appear in the density estimates and distribution function estimates.

1.1   Boundary Bias Problem

All of the previous explanations implicitly assume that the true density is supported on the entire real line. Howewer, if the data is nonnegatively supported, for instance, the standard kernel density estimator will suffer the so-called boundary bias problem. Throughout this section, the interval $[0, h]$ is called a "boundary region", and points greater than h are called "interior points".

In the boundary region, the standard kernel density estimator $\widehat{f}_X(x)$ usually underestimates $f_X(x)$. This is because it does not "feel" the boundary, and it puts weights for the lack of data on the negative axis. To be more precise, if a symmetric kernel supported on $[-1, 1]$ is used, then

$$Bias[\widehat{f}_X(x)] = \left[\int_{-1}^{c} K(u)du - 1\right] f_X(x) - hf_X'(x)\int_{-1}^{c} uK(u)du + O(h^2)$$

when $x \le h$, $c = \frac{x}{h}$. This means that this estimator is not consistent at $x = 0$ because

$$\lim_{n \to \infty} Bias[\widehat{f}_X(0)] = \left[\int_{-1}^{c} K(u)du - 1\right] f_X(0) \ne 0,$$

unless $f_X(0) = 0$.

Several ways of removing the boundary bias problem, each with their own advantages and disadvantages, are data reflection [8], simple nonnegative boundary correction [9], boundary kernels [10–12], pseudodata generation [13], a hybrid method [14], empirical transformation [15], a local linear estimator [16, 17], data binning and a local polynomial fitting on the bin counts [18], and others. Most of them use symmetric kernel functions as usual and then modify their forms or transform the data.

There is a simple way to circumvent the boundary bias that appears in the standard kernel density estimation. The remedy consists in replacing symmetric kernels with asymmetric gamma kernels, introduced by Chen [19], which never assign a weight outside of the support. Let $K(y; x, h)$ be an asymmetric function parameterized by x and h, called an "asymmetric kernel". Then, the definition of the asymmetric kernel density estimator is

$$\widehat{f}(x) = \frac{1}{n} \sum_{i=1}^{n} K(X_i; x, h). \tag{1.4}$$

Since the density of $\Gamma(xh^{-1} + 1, h)$, which is

$$\frac{y^{\frac{x}{h}} e^{-\frac{y}{h}}}{\Gamma\left(\frac{x}{h} + 1\right) h^{\frac{x}{h}+1}}, \tag{1.5}$$

is an asymmetric function parameterized by x and h, it is natural to use it as an asymmetric kernel. Hence, the gamma kernel density estimator is defined as

$$\widehat{f}_C(x) = \frac{1}{n} \sum_{i=1}^{n} \frac{X_i^{\frac{x}{h}} e^{-\frac{X_i}{h}}}{\Gamma\left(\frac{x}{h} + 1\right) h^{\frac{x}{h}+1}}. \tag{1.6}$$

The intuitive approach to seeing how Eq. (1.6) can be used as a consistent estimator is as follows. Let Y be a $\Gamma(xh^{-1} + 1, h)$ random variable with the pdf stated in Eq. (1.5), then

$$E[\widehat{f}_C(x)] = \int_{0}^{\infty} f_X(y) K(y; x, h) dy = E[f_X(Y)].$$

By Taylor expansion,

$$E[f_X(Y)] = f_X(x) + h\left[f_X'(x) + \frac{1}{2}xf_X''(x)\right] + o(h),$$

which will converge to $f_X(x)$ as $n \to \infty$. For a detailed theoretical explanation regarding the consistency of asymmetric kernels, see [20]. The bias and variance of Chen's gamma kernel density estimator are

$$Bias[\widehat{f_C}(x)] = \left[f_X'(x) + \frac{1}{2}xf_X''(x)\right]h + o(h), \tag{1.7}$$

$$Var[\widehat{f_C}(x)] = \begin{cases} \frac{f_X(x)}{2\sqrt{\pi}xn\sqrt{h}}, & \frac{x}{h} \to \infty, \\ \frac{\Gamma(2c+1)f_X(x)}{2^{2c+1}\Gamma^2(c+1)nh}, & \frac{x}{h} \to c, \end{cases} \tag{1.8}$$

for some $c > 0$.

Chen's gamma kernel density estimator obviously solved the boundary bias problem because the gamma pdf is a nonnegative supported function, so no weight will be put on the negative axis. However, it also has some problems, which are:

- the variance depends on a factor $x^{-1/2}$ in the interior, which means the variance becomes much larger quickly when x is small,
- the MSE is $O(n^{-2/3})$ when x is close to the boundary (worse than the standard kernel density estimator) [21].

1.2   Bias and Variance Reductions

This section discusses another kernel density estimator that has reduced bias and variance rates. Using a similar idea but with different parameters of gamma density as a kernel function, [22] intend to reduce the variance. Then the bias is reduced by modifying it with expansions of exponential and logarithmic functions. Hence, the modified gamma kernel density estimator is not only free of the boundary bias, but the variance also has smaller orders both in the interior and near the boundary, compared with Chen's method. As a result, the optimal orders of the MSE and the MISE are smaller as well.

Before continuing the discussion, some assumptions need to be imposed, which are:

A1. the bandwidth $h > 0$ satisfies $h \to 0$ and $nh \to \infty$ when $n \to \infty$,

A2. the density f_X is three times continuously differentiable, and $f_X^{(4)}$ exists,

A3. the following integrals $\int \left[\frac{f_X'(x)}{f_X(x)}\right]^2 dx$, $\int x^4 \left[\frac{f_X''(x)}{f_X(x)}\right]^2 dx$, $\int x^2[f_X''(x)]^2 dx$, and $\int x^6[f_X'''(x)]^2 dx$ are finite.

The first assumption is the usual assumption for the standard kernel density estimator. Since exponential and logarithmic expansions will be used, assumption A2 is needed to ensure the validity of the proofs. The last assumption is necessary to make sure the MISE can be calculated.

As stated before, the modification of the gamma kernel, done to improve the performance of Chen's method, is started by replacing the shape and scale parameters of the gamma density with suitable functions of x and h, and this kernel is defined as a new gamma kernel. The purpose in doing this is to reduce the variance so that it is smaller than the variance of Chen's method. After trying several combinations of functions, the density of $\Gamma(h^{-1/2}, x\sqrt{h} + h)$ is chosen, which is

$$K(y; x, h) = \frac{y^{\frac{1}{\sqrt{h}}-1} e^{-\frac{y}{x\sqrt{h}+h}}}{\Gamma\left(\frac{1}{\sqrt{h}}\right)(x\sqrt{h}+h)^{\frac{1}{\sqrt{h}}}},$$

as a kernel, and the new gamma kernel density "estimator" is defined as

$$A_h(x) = \frac{\sum_{i=1}^{n} X_i^{\frac{1}{\sqrt{h}}-1} e^{-\frac{X_i}{x\sqrt{h}+h}}}{n\Gamma\left(\frac{1}{\sqrt{h}}\right)(x\sqrt{h}+h)^{\frac{1}{\sqrt{h}}}}, \tag{1.9}$$

where n is the sample size, and h is the bandwidth.

Remark 1.1 Even though the formula in Eq. (1.9) can work as a density estimator properly, it is **not the proposed method**. As it will be stated later, another modification is needed for Eq. (1.9) before the proposed estimator is created.

Theorem 1.1 *Assuming A1 and A2, for the function $A_h(x)$ in Eq. (1.9), its bias and variance are*

$$Bias[A_h(x)] = \left[f_X'(x) + \frac{1}{2}x^2 f_X''(x)\right]\sqrt{h} + o(\sqrt{h}) \tag{1.10}$$

and

$$Var[A_h(x)] = \begin{cases} \dfrac{R^2\left(\frac{1}{\sqrt{h}}-1\right)f_X(x)}{2(x+\sqrt{h})\sqrt{\pi(1-\sqrt{h})}R\left(\frac{2}{\sqrt{h}}-2\right)nh^{\frac{1}{4}}} + O\left(\dfrac{h^{\frac{1}{4}}}{n}\right), & \frac{x}{h} \to \infty \\[2em] \dfrac{R^2\left(\frac{1}{\sqrt{h}}-1\right)f_X(x)}{2(c\sqrt{h}+1)\sqrt{\pi(1-\sqrt{h})}R\left(\frac{2}{\sqrt{h}}-2\right)nh^{\frac{3}{4}}} + O\left(\dfrac{1}{nh^{\frac{1}{4}}}\right), & \frac{x}{h} \to c, \end{cases} \tag{1.11}$$

for some positive number c, and

$$R(z) = \frac{\sqrt{2\pi}\, z^{z+\frac{1}{2}}}{e^z \Gamma(z+1)}. \tag{1.12}$$

Proof First, by usual reasoning of i.i.d. random variables, then

$$E[A_h(x)] = \int_0^\infty \frac{w^{\frac{1}{\sqrt{h}}-1}e^{-\frac{w}{x\sqrt{h}+h}}}{\Gamma\left(\frac{1}{\sqrt{h}}\right)(x\sqrt{h}+h)^{\frac{1}{\sqrt{h}}}} f_X(w)dw.$$

If $W \sim \Gamma(h^{-1/2}, x\sqrt{h}+h)$ with $\mu_W = h^{-1/2}(x\sqrt{h}+h)$, $Var(W) = h^{-1/2}(x\sqrt{h}+h)^2$, and $E[(W-\mu_W)^3] = 2h^{-1/2}(x\sqrt{h}+h)^3$, the integral can be seen as an expectation of $f_X(W)$. Hence, by Taylor expansion twice, first around μ_W, and next around x, result in

$$E[f_X(W)] = f_X(x) + \left[f_X'(x) + \frac{1}{2}x^2 f_X''(x)\right]\sqrt{h} + o(\sqrt{h}).$$

Hence, $Bias[A_h(x)]$ is in the order of $\sqrt{h}$.

Next is the derivation of the formula of the variance, which is

$$Var[A_h(x)] = n^{-1}E[K^2(X_1; x, h)] + O(n^{-1}).$$

First, the expectation part is

$$E[K^2(X_1; x, h)] = \frac{\Gamma\left(\frac{2}{\sqrt{h}}-1\right)\left(\frac{x\sqrt{h}+h}{2}\right)^{\frac{2}{\sqrt{h}}-1}}{\Gamma^2\left(\frac{1}{\sqrt{h}}\right)(x\sqrt{h}+h)^{\frac{2}{\sqrt{h}}}} \int_0^\infty \frac{v^{\left(\frac{2}{\sqrt{h}}-1\right)-1}e^{-\frac{2v}{x\sqrt{h}+h}}f_X(v)}{\Gamma\left(\frac{2}{\sqrt{h}}-1\right)\left(\frac{x\sqrt{h}+h}{2}\right)^{\frac{2}{\sqrt{h}}-1}}dv$$

$$= B(x, h)E[f_X(V)],$$

where V is a $\Gamma(2h^{-1/2}-1, (x\sqrt{h}+h)/2)$ random variable, $B(x, h)$ is a factor outside the integral, and the integral itself can be considered as $E[f_X(V)]$. In the same fashion as in $E[f_X(W)]$ before, it is clear that $E[f_X(V)] = f_X(x) + O(\sqrt{h})$. Now, let $R(z) = \frac{\sqrt{2\pi}z^{z+\frac{1}{2}}}{e^z\Gamma(z+1)}$, then $B(x, h)$ can be rewritten to become

$$B(x, h) = \frac{R^2\left(\frac{1}{\sqrt{h}}-1\right)}{2(x+\sqrt{h})\sqrt{\pi(1-\sqrt{h})}R\left(\frac{2}{\sqrt{h}}-2\right)h^{\frac{1}{4}}},$$

and the proof can be completed.

Remark 1.2 The function $R(z)$ monotonically increases with $\lim_{z\to\infty}R(z) = 1$ and $R(z) < 1$ [23], which means $\frac{R^2(\frac{1}{h}-1)}{R(\frac{2}{h}-2)} \le 1$. From these facts, it can be concluded that $Var[A_h(x)]$ is $O(n^{-1}h^{-1/4})$ when x is in the interior, and it is $O(n^{-1}h^{-3/4})$ when x is near the boundary. Both of these rates of convergence are faster than the rates of the variance of Chen's gamma kernel estimator for both cases, respectively. Furthermore, instead of $x^{-1/2}$, $Var[A_h(x)]$ depends on $(x + \sqrt{h})^{-1}$, which means the value of the variance will not speed up to infinity when x approaches 0.

Even though reducing the order of the variance is done, a larger bias order is encountered. To avoid this problem, geometric extrapolation to change the order of bias back to h can be utilized.

Theorem 1.2 *Let $A_h(x)$ be the function in Eq. (1.9). By assuming A1 and A2, also defining $J_h(x) = E[A_h(x)]$, then*

$$J_h^2(x)[J_{4h}(x)]^{-1} = f_X(x) + O(h). \tag{1.13}$$

Proof Now, extending the expansion of $J_h(x)$ until the h term results in

$$J_h(x) = f_X(x)\left[1 + \frac{a(x)}{f_X(x)}\sqrt{h} + \frac{b(x)}{f_X(x)}h + o(h)\right],$$

where $a(x) = f_X'(x) + \frac{1}{2}x^2 f_X''(x)$, and $b(x) = \left(x + \frac{1}{2}\right)f_X''(x) + x^2\left(\frac{x}{3} + \frac{1}{2}\right)f_X'''(x)$. By taking the natural logarithm and using its expansion, then

$$\log J_h(x) = \log f_X(x) + \frac{a(x)}{f_X(x)}\sqrt{h} + \left[b(x) - \frac{a^2(x)}{2f_X(x)}\right]\frac{h}{f_X(x)} + o(h).$$

Next, if $J_{4h}(x) = E[A_{4h}(x)]$ (using quadrupled bandwidth), i.e.,

$$\log J_{4h}(x) = \log f_X(x) + \frac{2a(x)}{f_X(x)}\sqrt{h} + \frac{4}{f_X(x)}\left[b(x) - \frac{a^2(x)}{2f_X(x)}\right]h + o(h),$$

it is possible to set up conditions to eliminate the term $\sqrt{h}$ while keeping the term $\log f_X(x)$. Now, since $\log[J_h(x)]^{t_1}[J_{4h}(x)]^{t_2}$ equals

$$(t_1 + t_2)\log f_X(x) + (t_1 + 2t_2)\frac{a(x)}{f_X(x)}\sqrt{h} + \frac{(t_1 + 4t_2)h}{f_X(x)}\left[b(x) - \frac{a^2(x)}{2f_X(x)}\right] + o(h),$$

the conditions needed are $t_1 + t_2 = 1$ and $t_1 + 2t_2 = 0$. It is obvious that the solution is $t_1 = 2$ and $t_2 = -1$, then

$$\log[J_h(x)]^2[J_{4h}(x)]^{-1} = \log f_X(x) - \frac{2}{f_X(x)}\left[b(x) - \frac{a^2(x)}{2f_X(x)}\right]h + o(h).$$

By taking the exponential function and using its expansion results in

$$[J_h(x)]^2[J_{4h}(x)]^{-1} = f_X(x) - 2\left[b(x) - \frac{a^2(x)}{2f_X(x)}\right]h + o(h).$$

$\square$

Remark 1.3 The function $J_{4h}(x)$ is the expectation of the function in Eq. (1.9) with $4h$ as the bandwidth. Furthermore, the term after $f_X(x)$ in Eq. (1.13) is in the order h, which is the same as the order of bias for Chen's gamma kernel density estimator.

Theorem 1.2 gives the idea to modify $A_h(x)$ and to define the new estimator. Hence,

$$\widetilde{f}_X(x) = [A_h(x)]^2[A_{4h}(x)]^{-1} \tag{1.14}$$

is proposed as the modified gamma kernel density estimator. This idea is actually straightforward. It uses the fact that the expectation of the operation of two statistics is asymptotically equal (in probability) to the operation of the expectation of each statistic.

For the bias of the proposed estimator, consult the following theorem.

Theorem 1.3 *Assuming A1 and A2, the bias of the modified gamma kernel density estimator is*

$$Bias[\widetilde{f}_X(x)] = -2\left[b(x) - \frac{a(x)}{2f_X(x)}\right]h + o(h) + O\left(\frac{1}{nh^{\frac{1}{4}}}\right), \tag{1.15}$$

where

$$a(x) = f_X'(x) + \frac{1}{2}x^2 f_X''(x), \tag{1.16}$$

$$b(x) = \left(x + \frac{1}{2}\right)f_X''(x) + x^2\left(\frac{x}{3} + \frac{1}{2}\right)f_X'''(x). \tag{1.17}$$

Proof By the definition, it can be rewritten that $A_h(x) = J_h(x) + Y$ and $A_{4h}(x) = J_{4h}(x) + Z$, where Y and Z are random variables with $E(Y) = E(Z) = 0$, $Var(Y) = Var[A_h(x)]$, and $Var(Z) = Var[A_{4h}(x)]$. Then, by $(1 + p)^q = 1 + pq + O(p^2)$, results in

$$\widetilde{f}_X(x) = [J_h(x)]^2[J_{4h}(x)]^{-1} + \frac{2J_h(x)}{J_{4h}(x)}Y - \left[\frac{J_h(x)}{J_{4h}(x)}\right]^2 Z + O[(Y + Z)^2].$$

Hence,

$$E[\widetilde{f}_X(x)] = f_X(x) - 2\left[b(x) - \frac{a(x)}{2f_X(x)}\right]h + o(h) + O\left(\frac{1}{nh^{\frac{1}{4}}}\right),$$

which leads to the desired result.

As expected, the bias' leading term is actually the same as the explicit form of $O(h)$ in Theorem 1.2. Its order of convergence changed back to h, the same as the bias of Chen's method. This is quite the accomplishment, because if keeping the order of the variance the same as $Var[A_h(x)]$ is possible, the MSE of the modified gamma kernel density estimator is smaller than the MSE of Chen's gamma kernel estimator. However, before jumping into the calculation of variance, the following theorem is needed.

Theorem 1.4 *Assuming A1 and A2, for the function in Eq. (1.9) with bandwidth h, $A_h(x)$, and with bandwidth 4h, $A_{4h}(x)$, the covariance of them is equal to*

$$Cov[A_h(x), A_{4h}(x)] = \frac{R\left(\frac{1}{\sqrt{h}} - 1\right) R\left(\frac{1}{2\sqrt{h}} - 1\right)}{2\sqrt{\pi} R\left(\frac{3}{2\sqrt{h}} - 2\right) (3x + 5\sqrt{h}) (2 - 2\sqrt{h})^{\frac{1}{\sqrt{h}} - \frac{1}{2}} (1 - 2\sqrt{h})^{\frac{1}{2\sqrt{h}} - \frac{1}{2}}} \frac{\left(\frac{3}{2} - 2\sqrt{h}\right)^{\frac{3}{2\sqrt{h}} - \frac{3}{2}}}{}$$

$$\times \left(\frac{x + \sqrt{h}}{3x + 5\sqrt{h}}\right)^{\frac{1}{2\sqrt{h}} - 1} \left(\frac{2x + 4\sqrt{h}}{3x + 5\sqrt{h}}\right)^{\frac{1}{\sqrt{h}} - 1} \frac{f_X(x)}{nh^{\frac{1}{4}}} + O\left(\frac{h^{\frac{1}{4}}}{n}\right),$$

when $xh^{-1} \to \infty$, and

$$Cov[A_h(x), A_{4h}(x)] = \frac{R\left(\frac{1}{\sqrt{h}} - 1\right) R\left(\frac{1}{2\sqrt{h}} - 1\right)}{2\sqrt{\pi} R\left(\frac{3}{2\sqrt{h}} - 2\right) (3c\sqrt{h} + 5) (2 - 2\sqrt{h})^{\frac{1}{\sqrt{h}} - \frac{1}{2}} (1 - 2\sqrt{h})^{\frac{1}{2\sqrt{h}} - \frac{1}{2}}} \frac{\left(\frac{3}{2} - 2\sqrt{h}\right)^{\frac{3}{2\sqrt{h}} - \frac{3}{2}}}{}$$

$$\times \left(\frac{c\sqrt{h} + 1}{3c\sqrt{h} + 5}\right)^{\frac{1}{2\sqrt{h}} - 1} \left(\frac{2c\sqrt{h} + 4}{3c\sqrt{h} + 5}\right)^{\frac{1}{\sqrt{h}} - 1} \frac{f_X(x)}{nh^{\frac{3}{4}}} + O\left(\frac{1}{nh^{\frac{1}{4}}}\right),$$

when $xh^{-1} \to c > 0$.

Proof By usual calculation of i.i.d. random variables, then

$$Cov[A_h(x), A_{4h}(x)] = \frac{1}{n} E[K(X_1; x, h) K(X_1; x, 4h)] + O\left(\frac{1}{n}\right).$$

Now, for the expectation,

$$E[K(X_1; x, h) K(X_1; x, 4h)] = \frac{\Gamma\left(\frac{3}{2\sqrt{h}} - 1\right) \left[\frac{2\sqrt{h}(x + \sqrt{h})(x + 2\sqrt{h})}{3x + 5\sqrt{h}}\right]^{\frac{3}{2\sqrt{h}} - 1}}{\Gamma\left(\frac{1}{\sqrt{h}}\right) \Gamma\left(\frac{1}{2\sqrt{h}}\right) (x\sqrt{h} + h)^{\frac{1}{\sqrt{h}}} (2x\sqrt{h} + 4h)^{\frac{1}{2\sqrt{h}}}}$$

$$\times \int_0^\infty \frac{t^{\left(\frac{3}{2\sqrt{h}} - 1\right) - 1} e^{-t\left[\frac{3x + 5\sqrt{h}}{2\sqrt{h}(x + \sqrt{h})(x + 2\sqrt{h})}\right]} f_X(t)}{\Gamma\left(\frac{3}{2\sqrt{h}} - 1\right) \left[\frac{2\sqrt{h}(x + \sqrt{h})(x + 2\sqrt{h})}{3x + 5\sqrt{h}}\right]^{\frac{3}{2\sqrt{h}} - 1}} dt$$

$$= C(x, h) E[f_X(T)],$$

where $C(x, h)$ is the factor outside the integral, and T is a random variable with

$$\mu_T = \frac{3(x + \sqrt{h})(x + 2\sqrt{h})}{3x + 5\sqrt{h}} + O(\sqrt{h})$$

and $Var(T) = O(\sqrt{h})$, which results in $E[f_X(T)] = f_X(x) + O(\sqrt{h})$. Using the definition of $R(z)$ as before results in

$$
C(x, h) = \frac{R\left(\frac{1}{\sqrt{h}} - 1\right) R\left(\frac{1}{2\sqrt{h}} - 1\right)}{2h^{\frac{1}{4}}\sqrt{\pi}R\left(\frac{3}{2\sqrt{h}} - 2\right)(3x + 5\sqrt{h})} \frac{\left(\frac{3}{2} - 2\sqrt{h}\right)^{\frac{3}{2\sqrt{h}} - \frac{3}{2}}}{(2 - 2\sqrt{h})^{\frac{1}{\sqrt{h}} - \frac{1}{2}}(1 - 2\sqrt{h})^{\frac{1}{2\sqrt{h}} - \frac{1}{2}}}
$$

$$
\times \left(\frac{x + \sqrt{h}}{3x + 5\sqrt{h}}\right)^{\frac{1}{2\sqrt{h}} - 1} \left(\frac{2x + 4\sqrt{h}}{3x + 5\sqrt{h}}\right)^{\frac{1}{\sqrt{h}} - 1},
$$

when $x > h$ (for $x \le h$, the calculation is similar). Hence, the theorem is proven. $\square$

Theorem 1.5 *Assuming A1 and A2, the variance of the modified gamma kernel density estimator is*

$$
Var[\widetilde{f}_X(x)] = 4Var[A_h(x)] + Var[A_{4h}(x)] - 4Cov[A_h(x), A_{4h}(x)] + o\left(\frac{1}{nh^{\frac{1}{4}}}\right),
$$

where its orders of convergence are $O(n^{-1}h^{-1/4})$ in the interior and $O(n^{-1}h^{-3/4})$ in the boundary region.

Proof It is easy to prove that $[J_h(x)][J_{4h}(x)]^{-1} = 1 + O(\sqrt{h})$ by using the expansion of $(1 + p)^q$. This fact brings to

$$
Var[\widetilde{f}_X(x)] = Var[2A_h(x) - A_{4h}(x)] + o\left(\frac{1}{nh^{\frac{1}{4}}}\right) \qquad (1.18)
$$

and the desired result. Lastly, since the result of Eq. (1.18) is just a linear combination of two variance formulas, the orders of the variance do not change, which are $n^{-1}h^{-1/4}$ in the interior and $n^{-1}h^{-3/4}$ in the boundary region. $\square$

As a conclusion to Theorems 1.3 and 1.5, with the identity of MSE, then

$$
MSE[\widetilde{f}_X(x)] = \begin{cases} O(h^2) + O\left(\frac{1}{nh^{\frac{1}{4}}}\right), & \frac{x}{h} \to \infty, \\ O(h^2) + O\left(\frac{1}{nh^{\frac{3}{4}}}\right), & \frac{x}{h} \to c. \end{cases} \qquad (1.19)
$$

The theoretical optimum bandwidths are $h = O(n^{-4/9})$ in the interior and $O(n^{-4/11})$ in the boundary region. As a result, the optimum orders of convergence are $O(n^{-8/9})$ and $O(n^{-8/11})$, respectively. Both of them are smaller than the optimum orders of Chen's estimator, which are $O(n^{-4/5})$ in the interior and $O(n^{-2/3})$ in the boundary region. Furthermore, since the MISE is just the integration of MSE, it is clear that the orders of convergence of the MISE are the same as those of the MSE.

Calculating the explicit formula of $MISE(\widetilde{f}_X)$ is nearly impossible because of the complexity of the formulas of $Bias[\widetilde{f}_X(x)]$ and $Var[\widetilde{f}_X(x)]$. However, by a similar argument stated in [19], the boundary region part of $Var[\widetilde{f}_X(x)]$ is negligible while integrating the variance. Thus, instead of computing $\int_{boundary} Var[\widetilde{f}_X(x)] + \int_{interior} Var[\widetilde{f}_X(x)]$, it is sufficient to just calculate $\int_0^\infty Var[\widetilde{f}_X(x)]dx$ using the formula of the variance in the interior. With that, computing

$$MISE(\widetilde{f}_X) = \int_0^\infty Bias^2[\widetilde{f}_X(x)]dx + \int_0^\infty Var[\widetilde{f}_X(x)]dx$$

can be approximated by numerical methods (assuming f_X is known).

1.3 Simulation Studies

Now, the results of a simulation study to show the performances of the proposed method and compare them with other estimators' results are provided. The measures of error used are the MISE, the MSE, bias, and variance. Since the assumptions A1, A2, and A3 are active, the MISE of the proposed estimator is finite. The average integrated squared error (AISE), the average squared error (ASE), simulated bias, and simulated variance are calculated, with the sample size of $n = 50$ and 10000 repetitions for each case.

Four gamma kernel density estimators are compared: Chen's gamma kernel density estimator $\widehat{f}_C(x)$, two nonnegative bias-reduced Chen's gamma estimators $\widehat{f}_{KI1}(x)$ and $\widehat{f}_{KI2}(x)$ [24], and the modified gamma kernel density estimator $\widetilde{f}_X(x)$. Several distributions for this study were generated, which are exponential distribution $exp(1/2)$, gamma distribution $\Gamma(2, 3)$, log-normal distribution $log.N(0, 1)$, Inverse Gaussian distribution $IG(1, 2)$, Weibull distribution $Wei(3, 2)$, and absolute normal distribution $abs.N(0, 1)$. The least squares cross-validation technique was used to determine the value of the bandwidths.

Table 1.1 compares AISEs, representing the general measure of error. The proposed method outperformed the other estimators. Since one of the main concerns is eliminating the boundary bias problem, it is necessary to take attention to the values of the measures of error in the boundary region. Tables 1.2–1.4 show the ASE, bias,

Table 1.1 Comparison of the average integrated squared error ($\times 10^5$)

Distributions	$\widehat{f}_C(x)$	$\widehat{f}_{KI1}(x)$	$\widehat{f}_{KI2}(x)$	$\widetilde{f}_X(x)$
$exp(1/2)$	970	1367	1304	**831**
$\Gamma(2, 3)$	313	2091	1913	**196**
$log.N(0, 1)$	342	1845	1688	**206**
$IG(1, 2)$	1002	680	660	**297**
$Wei(3, 2)$	7896	4198	4120	**1832**
$abs.N(0, 1)$	8211	3785	3719	**2905**

Table 1.2 Comparison of the average squared error ($\times 10^5$) when $x = 0.01$

Distributions	$\widehat{f_C}(x)$	$\widehat{f}_{K11}(x)$	$\widehat{f}_{K12}(x)$	$\widetilde{f}_X(x)$
$exp(1/2)$	1600	1547	1553	**991**
$\Gamma(2, 3)$	207	384	359	**168**
$log.N(0, 1)$	36	178	160	**34**
$IG(1, 2)$	1006	829	781	**422**
$Wei(3, 2)$	1528	708	643	**304**
$abs.N(0, 1)$	2389	2018	1999	**721**

Table 1.3 Comparison of the bias ($\times 10^4$) when $x = 0.01$

Distributions	$\widehat{f_C}(x)$	$\widehat{f}_{K11}(x)$	$\widehat{f}_{K12}(x)$	$\widetilde{f}_X(x)$
$exp(1/2)$	-1054	-1865	-1904	**-858**
$\Gamma(2, 3)$	391	583	561	**233**
$log.N(0, 1)$	150	417	395	**120**
$IG(1, 2)$	961	869	840	**386**
$Wei(3, 2)$	1215	821	780	**342**
$abs.N(0, 1)$	-1383	303	297	**157**

Table 1.4 Comparison of the variance ($\times 10^5$) when $x = 0.01$

Distributions	$\widehat{f_C}(x)$	$\widehat{f}_{K11}(x)$	$\widehat{f}_{K12}(x)$	$\widetilde{f}_X(x)$
$exp(1/2)$	490	1465	1469	**244**
$\Gamma(2, 3)$	54	43	44	**11**
$log.N(0, 1)$	39	36	36	**35**
$IG(1, 2)$	835	739	753	**273**
$Wei(3, 2)$	532	340	343	**184**
$abs.N(0, 1)$	476	1926	1910	**211**

and variance of those four estimators when $x = 0.01$. Once again, the proposed estimator had the best results. Though the differences among the values of bias were relatively not big (Table 1.3), Table 1.4 illustrates how the variance reduction has an effect.

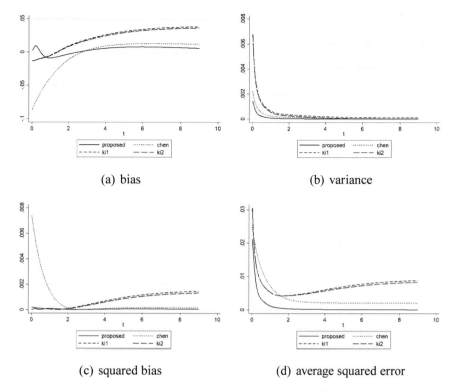

(a) bias

(b) variance

(c) squared bias

(d) average squared error

Fig. 1.1 Comparison of the point-wise bias, variance, and ASE of $\tilde{f}_X(x)$, $\hat{f}_C(x)$, $\hat{f}_{KI1}(x)$, and $\hat{f}_{KI2}(x)$ for estimating density of $\Gamma(2, 3)$ with sample size $n = 150$

As further illustrations, graphs of point-wise ASE, bias, squared bias, and variance to compare the proposed estimator's performances with those of the others are also provided. The exponential, gamma, and absolute normal were generated 1000 times to produce Figs. 1.2–1.3.

In some cases, the bias value of the proposed estimator was away from 0 more than the other estimators (e.g., Fig. 1.2a around $x = 1$, Fig. 1.1a around $x = 4$, and Fig. 1.3a around $x = 0.2$). Though this could reflect poorly on the proposed estimator, from the variance parts (Figs. 1.2b, 1.1b, and 1.3b), the estimator never failed to give the smallest value of variance, confirming the successfulness of the method of

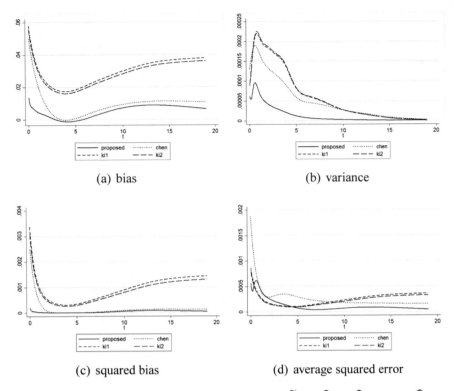

(a) bias   (b) variance

(c) squared bias   (d) average squared error

Fig. 1.2 Comparison of point-wise bias, variance, and ASE of $\widetilde{f}_X(x)$, $\widehat{f}_C(x)$, $\widehat{f}_{KI1}(x)$, and $\widehat{f}_{KI2}(x)$ for estimating density of $\exp(1/2)$ with sample size $n = 150$

variance reduction proposed here. Moreover, the result of the variance reduction is the reduction of point-wise ASE itself, shown in Figs. 1.2d, 1.1d, and 1.3d. One may take note of Fig. 1.1d when $x \in [1, 4]$ because the estimators from [24] slightly outperformed the proposed method. However, as x got larger, $ASE[\widehat{f}_{KI1}(x)]$ and $ASE[\widehat{f}_{KI2}(x)]$ failed to get closer to 0 (they will when x is large enough), while $ASE[\widetilde{f}_X(x)]$ approached 0 immediately.

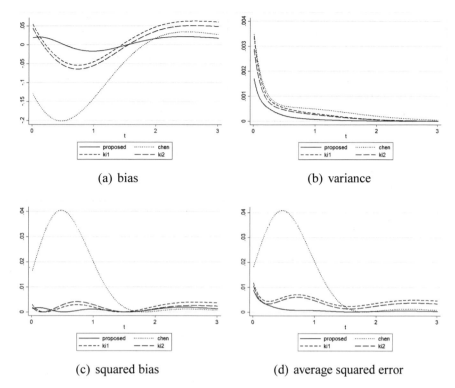

(a) bias

(b) variance

(c) squared bias

(d) average squared error

Fig. 1.3 Comparison of the point-wise bias, variance, and ASE of $\tilde{f}_X(x)$, $\hat{f}_C(x)$, $\hat{f}_{KI1}(x)$, and $\hat{f}_{KI2}(x)$ for estimating density of $abs.N(0, 1)$ with sample size $n = 150$

References

1. Parzen E (1962) On estimation of a probability density function and mode. Ann Math Stat 32:1065–1076
2. Rosenblatt M (1956) Remarks on some non-prametric estimates of a density function. Ann Math Stat 27:832–837
3. Terrel GR, Scott DW (1980) On improving convergence rates for non-negative kernel density estimation. Ann Stat 8:1160–1163
4. Abramson IS (1982) On bandwidth variation in kernel estimates-a square root law. Ann Stat 10:1217–1223
5. Samiuddin M, El-Sayyad GM (1990) On nonparametric kernel density estimates. Biom 77:865–874
6. Ruppert DR, Cline DBH (1994) Bias reduction in kernel density estimation by smoothed empirical transformations. Ann Stat 22:185–210
7. Jones MC, Linton O, Nielsen JP (1995) A simple bias reduction method for density estimation. Biom 82:327–338
8. Schuster EF (1985) Incorporating support constraints into nonparametric estimators of densities. Commun Stat Theory Method 14:1123–1136
9. Jones MC, Foster PJ (1996) A simple nonnegative boundary correction method for kernel density estimation. Stat Sin 6:1005–1013

10. Müller HG (1991) Smooth optimum kernel estimators near endpoints. Biom 78:521–530
11. Müller HG (1993) On the boundary kernel method for nonparametric curve estimation near endpoints. Scand J Stat 20:313–328
12. Müller HG, Wang JL (1994) Hazard rate estimation under random censoring with varying kernels and bandwidths. Biom 50:61–76
13. Cowling A, Hall P (1996) On pseudodata methods for removing boundary effects in kernel density estimation. J R Stat Soc B 58:551–563
14. Hall P, Wehrly TE (1991) A geometrical method for removing edge effects from kernel-type nonparametric regression estimators. J Am Stat Assoc 86:665–672
15. Marron JS, Ruppert D (1994) Transformation to reduce boundary bias in kernel density estimation. J R Stat Soc B 56:653–671
16. Lejeune M, Sarda P (1992) Smooth estimators of distribution and density functions. Comput Stat Data Anal 14:457–471
17. Jones MC (1993) Simple boundary correction for kernel density estimation. Stat Comput 3:135–146
18. Cheng MY, Fan J, Marron JS (1997) On automatic boundary corrections. Ann Stat 25:1691–1708
19. Chen SX (2000) Probability density function estimation using gamma kernels. Ann Inst Stat Math 52:471–480
20. Bouezmarni T, Scaillet O (2005) Consistency of asymmetric kernel density estimators and smoothed histograms with application to income data. Econ Theory 21:390–412
21. Zhang S (2010) A note on the performance of the gamma kernel estimators at the boundary. Stat Probab Lett 80:548–557
22. Fauzi RR, Maesono Y (2020) New type of gamma kernel density estimator. J Korean Stat Soc 49:882–900
23. Brown BM, Chen SX (1999) Beta-Bernstein smoothing for regression curves with compact supports. Scand J Stat 26:47–59
24. Igarashi G, Kakizawa Y (2015) Bias corrections for some asymmetric kernel estimators. J Stat Plan Inference 159:37–63

Chapter 2
Kernel Distribution Function Estimator

Abstract After discussing density estimation, here the estimation of distribution function is studied. Started from the standard distribution function estimator and its properties, a method to reduce the mean integrated squared error for kernel distribution function estimators is also proposed. It can be shown that the asymptotic bias of the proposed method is considerably smaller in the sense of convergence rate than that of the standard one, and even the variance of the proposed method is smaller up to some constants. The idea of this method is using a self-elimination technique between two standard kernel distribution function estimators with different bandwidths, with some helps of exponential and logarithmic expansions. As a result, the mean squared error can be reduced.

In statistical inference, statistics can be regarded as functions of the empirical distribution function. Since the empirical distribution function is not smooth, it is quite natural to replace it with the kernel-type distribution function. If the kernel-type distribution estimator is used, it results in smooth estimator and may reduce mean squared errors. Especially the class of L-statistics which is a linear combination of the order statistics is a functional of the empirical distribution function and so it is easy to extend the L-statistics based on the kernel type distribution estimator. In this chapter, the properties of the estimators of interesting parameters which are based on kernel estimator will be studied.

Let $X_1, X_2, \ldots, X_n$ be independently and identically distributed random variables with an absolutely continuous distribution function F_X and a density f_X. The classical nonparametric estimator of F_X has been the empirical distribution function defined by

$$F_n(x) = \frac{1}{n} \sum_{i=1}^{n} I(X_i \leq x), \quad x \in \mathbb{R}, \tag{2.1}$$

where $I(A)$ denotes the indicator function of a set A. It is obvious that F_n is a step function of height $\frac{1}{n}$ at each observed sample point X_i. When considered as a pointwise estimator of F_X, $F_n(x)$ is an unbiased and strongly consistent estimator of

© The Author(s), under exclusive license to Springer Nature Singapore Pte Ltd. 2023
Rizky Reza Fauzi and Y. Maesono, *Statistical Inference Based on Kernel Distribution Function Estimators*, JSS Research Series in Statistics,
https://doi.org/10.1007/978-981-99-1862-1_2

$F_X(x)$. From the global point of view, the Glivenko-Cantelli Theorem implies that

$$\sup \left\{|F_n(x) - F_X(x)|\big|x \in \mathbb{R}\right\} \to 0 \quad \text{a.s.}$$

as $n \to \infty$. For details, see Sect. 2.1 of [1]. However, given the information that F_X is absolutely continuous, it seems to be more appropriate to use a smooth and continuous estimator of F_X rather than the empirical distribution function F_n.

However, given the information that F_X is absolutely continuous, it seems to be more appropriate to use a smooth and continuous estimator of F_X rather than the empirical distribution function F_n. Nadaraya [2] defined it as

$$\widehat{F}_h(x) = \frac{1}{n} \sum_{i=1}^{n} W\left(\frac{x - X_i}{h}\right), \quad x \in \mathbb{R}, \tag{2.2}$$

where $W(v) = \int_{-\infty}^{v} K(w)\mathrm{d}w$. It is easy to prove that this kernel distribution function estimator is continuous, and satisfies all the properties of a distribution function.

2.1   Properties of KDFE

Several properties of $\widehat{F}_h(x)$ are well known. The almost sure uniform convergence of $\widehat{F}_h$ to F_X was proved in [2–4], while [5] extended this result to higher dimensions. The asymptotic normality of $\widehat{F}_h(x)$ was proven in [6], and Chung-Smirnov Property was established by [7, 8], i.e.,

$$\limsup_{n \to \infty} \sqrt{\frac{2n}{\log \log n}} \sup \left\{|\widehat{F}_h(x) - F_X(x)|\big|x \in \mathbb{R}\right\} = 1 \quad \text{a.s..}$$

Moreover, several authors showed that the asymptotic performance of $\widehat{F}_h(x)$ is better than that of $F_n(x)$, see [9–16].

Under the condition that f_X (the density) has one continuous derivative f_X', it has been proved by the above-mentioned authors that, as $n \to \infty$,

$$Bias[\widehat{F}_h(x)] = h^2 \frac{f_X'(x)}{2} \int_{-\infty}^{\infty} z^2 K(z)\mathrm{d}z + o(h^2), \tag{2.3}$$

$$Var[\widehat{F}_h(x)] = \frac{1}{n} F_X(x)[1 - F_X(x)] - \frac{2h}{n} r_1 f_X(x) + o\left(\frac{h}{n}\right), \tag{2.4}$$

where $r_1 = \int_{-\infty}^{\infty} y K(y) W(y)\mathrm{d}y$. If K is a symmetric kernel, then r_1 is a nonnegative number. Using the fact $K(-x) = K(x)$ and $W(-u) = 1 - W(u)$ means

$$r_1 = \int_{-\infty}^{0} yK(y)W(y)dy + \int_{0}^{\infty} yK(y)W(y)dy$$

$$= -\int_{0}^{\infty} uK(u)W(-u)du + \int_{0}^{\infty} yK(y)W(y)dy$$

$$= \int_{0}^{\infty} uK(u)\{2W(u) - 1\}du.$$

Since $K(u) \geq 0$ and $W(u) \geq \frac{1}{2}$ ($u \geq 0$), then $r_1 > 0$. If an appropriate higher order kernel is chosen, it can be concluded that $r_1 > 0$. The bias and the variance of the empirical distribution function are

$$Bias[F_n(x)] = 0,$$
$$Var[F_n(x)] = \frac{1}{n}F_X(x)[1 - F_X(x)]$$

Therefore, if a kernel which satisfies $r_1 > 0$ is chosen, the kernel distribution estimator is superior to the empirical distribution for any F_X.

Because of the relationship

$$MISE(\widehat{F}_h) = \int_{-\infty}^{\infty} [Bias^2\{\widehat{F}_h(x)\} + Var\{\widehat{F}_h(x)\}]dx,$$

then it is easy to conclude that

$$MISE(\widehat{F}_h) = \frac{h^4}{4}\left[\int_{-\infty}^{\infty} z^2 K(z)dz\right]^2 \int_{-\infty}^{\infty} [f_X'(x)]^2 dx$$
$$+ \frac{1}{n}\int_{-\infty}^{\infty} F_X(x)[1 - F_X(x)]dx - \frac{2h}{n}r_1 + o\left(h^4 + \frac{h}{n}\right)$$

provided all the integrals above are finite.

2.2 Bias Reduction of KDFE

There have been many proposals in the literature for improving the bias of the standard kernel density estimator. Typically, under sufficient smoothness conditions placed on the underlying density f_X, the bias reduces from $O(h^2)$ to $O(h^4)$, and the variance remains of order $\frac{1}{nh}$. Because of the good performances of the method of geometric extrapolation in [3] for density estimator, the similar idea is used in [17] to improve the standard kernel distribution function estimator. However, instead of a fixed multiplication factor for the bandwidth, a general term for that is used. It can be shown that the proposed estimator, $\tilde{F}_X$, has a smaller bias in the sense of convergence rate, that is $O(h^4)$. Furthermore, even though the rate of convergence of variance does not change, the variance of the proposed method is smaller up to some constants. At last, it has improved MISE as well.

The idea of reducing bias by geometric extrapolation is doing a self-elimination technique between two standard kernel distribution function estimators with different bandwidths, with some help of exponential and logarithmic expansions. By doing that, vanishing the h^2 term of the asymptotic bias is possible, and the order of convergence changes to h^4.

Before starting the main purpose, some assumptions are imposed, they are:

B1. the kernel K is a nonnegative continuous function, symmetric about 0, and it integrates to 1,
B2. the integral $\int_{-\infty}^{\infty} w^4 K(x)\mathrm{d}w$ is finite,
B3. the bandwidth $h > 0$ satisfies $h \to 0$ and $nh \to \infty$ when $n \to \infty$,
B4. the density f_X is three times continuously differentiable, and $f_X^{(4)}$ exists,
B5. the integrals $\int_{-\infty}^{\infty} \frac{[f_X'(x)]^2}{F_X(x)}\mathrm{d}x$ and $\int_{-\infty}^{\infty} f_X'''(x)\mathrm{d}x$ are finite.

The first and third assumptions are the usual assumptions for the standard kernel distribution function estimator. Since the exponential and logarithmic expansions are used, assumptions B2 and B4 are needed to ensure the validity of the proofs. For the last assumption, it is necessary to make sure the MISE can be calculated.

Theorem 2.1 *Let $J_h(x) = E[\widehat{F}_h(x)]$ and $a(\neq 1)$ be a positive number. Under the assumptions B1–B4, then*

$$[J_h(x)]^{t_1}[J_{ah}(x)]^{t_2} = F_X(x) + O(h^4),$$

where $t_1 = \frac{a^2}{a^2-1}$ and $t_2 = -\frac{1}{a^2-1}$.

Proof Let $J_h(x) = E[\widehat{F}_h(x)]$, and extend the expansion until h^4 term, which is

$$J_h(x) = W\left(\frac{x-v}{h}\right) F_X(v)\Big|_{-\infty}^{\infty} + \frac{1}{h}\int_{-\infty}^{\infty} F_X(v)H\left(\frac{x-v}{h}\right)\mathrm{d}v$$

$$= F_X(x)\left[1 + h^2\frac{b_2(x)}{F_X(x)} + h^4\frac{b_4(x)}{F_X(x)} + o(h^4)\right],$$

where $b_2(x) = \frac{f_X'(x)}{2}\int_{-\infty}^{\infty} w^2 K(w)\mathrm{d}w$ and $b_4(x) = \frac{f_X'''(x)}{24}\int_{-\infty}^{\infty} w^4 K(w)\mathrm{d}w$. By taking a natural logarithm and using its expansion, then

$$\log J_h(x) = \log F_X(x) + \log\left[1 + h^2\frac{b_2(x)}{F_X(x)} + h^4\frac{b_4(x)}{F_X(x)} + o(h^4)\right]$$

$$= \log F_X(x) + \sum_{k=1}^{\infty} \frac{(-1)^{k-1}}{k}\left[h^2\frac{b_2(x)}{F_X(x)} + h^4\frac{b_4(x)}{F_X(x)} + o(h^4)\right]^k$$

$$= \log F_X(x) + h^2\frac{b_2(x)}{F_X(x)} + h^4\frac{2b_4(x)F_X(x) - b_2^2(x)}{2F_X^2(x)} + o(h^4).$$

Next, if $J_{ah}(x) = E[\widehat{F}_{ah}(x)]$, which means

$$\log J_{ah}(x) = \log F_X(x) + a^2 h^2 \frac{b_2(x)}{F_X(x)} + a^4 h^4 \frac{2b_4(x) F_X(x) - b_2^2(x)}{2 F_X^2(x)} + o(h^4),$$

some conditions to eliminate the term h^2 but keep the term $\log F_X(x)$ can be set up. Since $\log[J_h(x)]^{t_1}[J_{ah}(x)]^{t_2}$ equals

$$(t_1 + t_2) \log F_X(x) + (t_1 + a^2 t_2)h^2 \frac{b_2(x)}{F_X(x)}$$

$$+ (t_1 + a^4 t_2)h^4 \frac{2b_4(x) F_X(x) - b_2^2(x)}{2 F_X^2(x)} + o(h^4),$$

the conditions needed are $t_1 + t_2 = 1$ and $t_1 + a^2 t_2 = 0$. It is obvious that the solutions are $t_1 = \frac{a^2}{a^2-1}$ and $t_2 = -\frac{1}{a^2-1}$, and then

$$\log[J_h(x)]^{\frac{a^2}{a^2-1}}[J_{ah}(x)]^{-\frac{1}{a^2-1}} = \log F_X(x) - h^4 a^2 \frac{2b_4(x) F_X(x) - b_2^2(x)}{2 F_X^2(x)} + o(h^4).$$

Using exponential function and its expansion to neutralize the natural logarithmic function, then

$$[J_h(x)]^{\frac{a^2}{a^2-1}}[J_{ah}(x)]^{-\frac{1}{a^2-1}}$$

$$= F_X(x) \sum_{k=0}^{\infty} \frac{(-1)^k}{k!} \left[h^4 a^2 \frac{2b_4(x) F_X(x) - b_2^2(x)}{2 F_X^2(x)} + o(h^4) \right]^k$$

$$= F_X(x) + h^4 a^2 \frac{b_2^2(x) - 2b_4(x) F_X(x)}{2 F_X(x)} + o(h^4).$$

Remark 2.1 The function $J_{ah}(x)$ is an expectation of the standard kernel distribution function estimator with ah as the bandwidth, that is, $J_{ah}(x) = E[\widehat{F}_{ah}(x)]$, where

$$\widehat{F}_{ah}(x) = \frac{1}{n} \sum_{i=1}^{n} W\left(\frac{x - X_i}{ah}\right), \quad x \in \mathbb{R}.$$

Furthermore, the term after $F_X(x)$ in the expansion is in the order h^4, which is smaller than the order of bias of the standard kernel distribution function estimator. Even though this theorem does not state a bias of some estimators, it will lead to the idea to modify the standard kernel distribution function estimator.

Theorem 2.1 gives an idea to modify kernel distribution function estimator which will have, intuitively, similar property for bias. Hence, the proposed estimator of distribution function is

$$\tilde{F}_X(x) = [\widehat{F}_h(x)]^{\frac{a^2}{a^2-1}}[\widehat{F}_{ah}(x)]^{-\frac{1}{a^2-1}}. \tag{2.5}$$

This idea is actually straightforward, using the fact that the expectation of the operation of two statistics is asymptotically equal (in probability) to the operation of expectation of each statistic.

Remark 2.2 The number a acts as the second smoothing parameter here, because it controls the smoothness of $\widehat{F}_{ah}$ (since it is placed inside the function W), and determines how much the effect of $\widehat{F}_h$ and $\widehat{F}_{ah}$ as a part of their power. Larger a means the effect of $\widehat{F}_h$ is larger for $\tilde{F}_X$, and vice versa. Furthermore, when $a \to \infty$, it can be seen that $\tilde{F}_X \to \widehat{F}_h$. Oppositely, when a is really close to 0, the effect of $\widehat{F}_h$ has almost vanished. However, different with bandwidth h, the number a can be chosen freely and does not depend on the sample size n. Letting a too close to 0 is not wise, since it acts as a denominator in the argument of function W.

Theorem 2.2 *Under the assumptions B1–B4, the bias of $\tilde{F}_X(x)$ is given by*

$$Bias[\tilde{F}_X(x)] = h^4 a^2 \frac{b_2^2(x) - 2b_4(x)F_X(x)}{2F_X(x)} + o(h^4) + O\left(\frac{1}{n}\right), \tag{2.6}$$

where

$$b_2(x) = \frac{f_X'(x)}{2} \int_{-\infty}^{\infty} w^2 K(w) dw, \tag{2.7}$$

$$b_4(x) = \frac{f_X'''(x)}{24} \int_{-\infty}^{\infty} w^4 K(w) dw. \tag{2.8}$$

Proof In order to investigate the bias of the proposed estimator, consider

$$\widehat{F}_h(x) = J_h(x) + Y$$

and $\widehat{F}_{ah}(x) = J_{ah}(x) + Z$, where Y and Z are random variables with

$$E(Y) = E(Z) = 0,$$

$Var(Y) = Var[\widehat{F}_h(x)]$, and $Var(Z) = Var[\widehat{F}_{ah}(x)]$. These forms are actually reasonable, because of the definition of $J_h(x)$ and $J_{ah}(x)$. Then, by the expansion

$$(1 + p)^q = 1 + pq + O(p^2),$$

then

$$\tilde{F}_X(x) = [J_h(x)]^{\frac{a^2}{a^2-1}}[J_{ah}(x)]^{-\frac{1}{a^2-1}}\left[1 + \frac{Y}{J_h(x)}\right]^{\frac{a^2}{a^2-1}}\left[1 + \frac{Z}{J_{ah}(x)}\right]^{-\frac{1}{a^2-1}}$$

$$= [J_h(x)]^{\frac{a^2}{a^2-1}}[J_{ah}(x)]^{-\frac{1}{a^2-1}}\left[1 + \frac{a^2}{a^2-1}\frac{Y}{J_h(x)} + O\left\{\frac{Y^2}{J_h^2(x)}\right\}\right]$$

$$\times\left[1 - \frac{1}{a^2-1}\frac{Z}{J_{ah}(x)} + O\left\{\frac{Z^2}{J_{ah}^2(x)}\right\}\right]$$

$$= [J_h(x)]^{\frac{a^2}{a^2-1}}[J_{ah}(x)]^{-\frac{1}{a^2-1}} + \frac{a^2}{a^2-1}\left[\frac{J_h(x)}{J_{ah}(x)}\right]^{\frac{1}{a^2-1}}Y$$

$$-\frac{1}{a^2-1}\left[\frac{J_h(x)}{J_{ah}(x)}\right]^{\frac{a^2}{a^2-1}}Z + O[(Y+Z)^2].$$

Hence,

$$E[\tilde{F}_X(x)] = E\left[\{J_h(x)\}^{\frac{a^2}{a^2-1}}\{J_{ah}(x)\}^{-\frac{1}{a^2-1}} + \frac{a^2}{a^2-1}\left\{\frac{J_h(x)}{J_{ah}(x)}\right\}^{\frac{1}{a^2-1}}Y\right.$$

$$\left.-\frac{1}{a^2-1}\left\{\frac{J_h(x)}{J_{ah}(x)}\right\}^{\frac{a^2}{a^2-1}}Z + O\{(Y+Z)^2\}\right]$$

$$= F_X(x) + h^4 a^2\frac{b_2^2(x) - 2b_4(x)F_X(x)}{2F_X(x)} + o(h^4) + O\left(\frac{1}{n}\right),$$

and the bias formula is derived. □

Remark 2.3 The factor $\frac{[f_X'(x)]^2}{F_X(x)}$ gives some uncertain feelings that this bias may be unbounded in some points of real line, especially when $x \to -\infty$. However, even though it is not stated in the theorem, assumption B5 ensures that the bias is valid and bounded a.s. on the real line. For a brief example, when the true distribution is the standard normal distribution with $f_X = \phi$, simple L'Hôpital's rule results in

$$\lim_{x\to-\infty}\frac{[f_X'(x)]^2}{F_X(x)} = \lim_{x\to-\infty}\frac{2x(1-x^2)\phi^2(x)}{\phi(x)} = 0.$$

Remark 2.4 As expected, the bias is in the order of h^4. This order is same as if the forth order kernel function for the standard kernel distribution function estimator was used. However, the proposed estimator is more appealing as it never has negative value.

Next, the property of variance would be discussed. Interestingly enough, there are no differences between the variance of the proposed estimator and the variance of the standard kernel distribution function estimator, in the sense of convergence order, as stated in the theorem below.

Theorem 2.3 *Under the assumptions B1–B4, the variance of $\tilde{F}_X(x)$ is given by*

$$Var[\tilde{F}_X(x)] = \frac{1}{n}F_X(x)[1 - F_X(x)] - \frac{h}{n}\left[\frac{2(a^4 + 1)}{(a^2 - 1)^2}r_1 + r_2\right]f_X(x) + o\left(\frac{h}{n}\right),$$

where

$$r_1 = \int_{-\infty}^{\infty} yK(y)W(y)dy, \tag{2.9}$$

$$r_2 = \int_{-\infty}^{\infty} y\left[K(y)W\left(\frac{y}{a}\right) + \frac{1}{a}W(y)K\left(\frac{y}{a}\right)\right]dy. \tag{2.10}$$

Proof Using $(1 + p)^q = 1 + pq + O(p^2)$, then

$$\frac{J_h(x)}{J_{ah}(x)} = \frac{1 + h^2\frac{b_2(x)}{F_X(x)} + h^4\frac{b_4(x)}{F_X(x)} + o(h^4)}{1 + O(h^2)}$$

$$= \left[1 + h^2\frac{b_2(x)}{F_X(x)} + h^4\frac{b_4(x)}{F_X(x)} + o(h^4)\right][1 + O(h^2) + O(h^4)]$$

$$= 1 + h^2\frac{b_2(x)}{F_X(x)} + h^4\frac{b_4(x)}{F_X(x)} + o(h^4) + O(h^2) = 1 + O(h^2).$$

The calculation of the variance is

$$Var[\tilde{F}_X(x)] = Var\left(\frac{a^2}{a^2 - 1}Y - \frac{1}{a^2 - 1}Z\right)[1 + O(h^2)] + O\left(\frac{1}{n^2}\right)$$

$$= Var\left(\frac{a^2}{a^2 - 1}Y - \frac{1}{a^2 - 1}Z\right) + O\left(\frac{h^2}{n} + \frac{1}{n^2}\right)$$

$$= Var\left[\frac{a^2}{a^2 - 1}\widehat{F}_h(x) - \frac{1}{a^2 - 1}\widehat{F}_{ah}(x)\right] + o\left(\frac{h}{n}\right).$$

Because this is just a variance of linear combination of two standard kernel distribution function estimators, the order of the variance does not change, that is n^{-1}. For the explicit formula of the variance, first, calculate

$$\frac{a^4}{(a^2 - 1)^2}Var[\widehat{F}_h(x)] + \frac{1}{(a^2 - 1)^2}Var[\widehat{F}_{ah}(x)] - \frac{2a^2}{(a^2 - 1)^2}Cov[\widehat{F}_h(x), \widehat{F}_{ah}(x)].$$

Since the formulas of $Var[\widehat{F}_h(x)]$ and $Var[\widehat{F}_{ah}(x)]$ are known, considering the covariance part is enough, which is

$$Cov[\widehat{F}_h(x), \widehat{F}_{ah}(x)] = \frac{1}{n^2} \sum_{i=1}^{n} \sum_{j=1}^{n} Cov\left[W\left(\frac{x-X_i}{h}\right), W\left(\frac{x-X_j}{ah}\right)\right]$$

$$= \frac{1}{n} Cov\left[W\left(\frac{x-X_1}{h}\right), W\left(\frac{x-X_1}{ah}\right)\right]$$

$$= \frac{1}{n}\left[E\left\{W\left(\frac{x-X_1}{h}\right) W\left(\frac{x-X_1}{ah}\right)\right\}\right.$$

$$\left. - E\left\{W\left(\frac{x-X_1}{h}\right)\right\} E\left\{W\left(\frac{x-X_1}{ah}\right)\right\}\right].$$

Because,

$$E\left[W\left(\frac{x-X_1}{h}\right)\right] - E\left[W\left(\frac{x-X_1}{ah}\right)\right] = F_X(x) + O(h^2),$$

then

$$E\left[W\left(\frac{x-X_1}{h}\right) W\left(\frac{x-X_1}{ah}\right)\right]$$

$$= \frac{1}{h} \int_{-\infty}^{\infty} F_X(x)\left[K\left(\frac{x-v}{h}\right) W\left(\frac{x-v}{ah}\right) + \frac{1}{a}W\left(\frac{x-v}{h}\right) K\left(\frac{x-v}{ah}\right)\right] dv$$

$$= \int_{-\infty}^{\infty} [F_X(x) - hyf_X(x) + o(h)]\left[K(y)W\left(\frac{y}{a}\right) + \frac{1}{a}W(y)K\left(\frac{y}{a}\right)\right] dy$$

$$= F_X(x)\left[\int_{-\infty}^{\infty} K(y)W\left(\frac{y}{a}\right) dy + \frac{1}{a}\int_{-\infty}^{\infty} W(y)K\left(\frac{y}{a}\right) dy\right] - hf_X(x)r_2 + o(h),$$

where $r_2 = \int_{-\infty}^{\infty} y\left[K(y)W\left(\frac{y}{a}\right) + \frac{1}{a}W(y)K\left(\frac{y}{a}\right)\right] dy$. For the first term of the right-hand side, hence

$$\int_{-\infty}^{\infty} K(y)W\left(\frac{y}{a}\right) dy = W\left(\frac{y}{a}\right) W(y)\Big|_{-\infty}^{\infty} - \frac{1}{a}\int_{-\infty}^{\infty} W(y)K\left(\frac{y}{a}\right) dy$$

$$= 1 - \frac{1}{a}\int_{-\infty}^{\infty} W(y)K\left(\frac{y}{a}\right) dy.$$

Thus,

$$E\left[W\left(\frac{x-X_1}{h}\right) W\left(\frac{x-X_1}{ah}\right)\right] = F_X(x) - hf_X(x)r_2 + o(h).$$

As a result, it can be shown that

$$Cov[\widehat{F}_h(x), \widehat{F}_{ah}(x)] = \frac{1}{n}F_X(x)[1 - F_X(x)] - \frac{h}{n}f_X(x)r_2 + o\left(\frac{h}{n}\right),$$

and the theorem is proven.

Remark 2.5 Actually in many cases, the hn^{-1} term is omitted and just denote it as $O(hn^{-1})$. However, since the dominant term of the variance of the standard kernel distribution function estimator and the proposed method are same, the second-order term is needed to compare them. It is easy to show that $\frac{a^2}{(a^2-1)^2}r_2 \geq 0$ and $\frac{a^4+1}{(a^2-1)^2} \geq 1$ when $a < 1$ (which is suggested). Hence, up to some constants, the variance of the proposed estimator is smaller than the standard kernel distribution function estimator's variance.

Since both the bias and the variance of the proposed estimator are smaller, then the MISE of the proposed estimator is smaller than the MISE of the standard kernel distribution function estimator. The following theorem states that clearly.

Theorem 2.4 *Under assumptions B1–B5, the mean integrated square error of* $\tilde{F}_X$ *is smaller than the MISE of* $\widehat{F}_h$. *It is given by*

$$MISE(\tilde{F}_X) = h^8 a^4 \int_{-\infty}^{\infty} \left[\frac{b_2^2(x) - 2b_4(x)F_X(x)}{2F_X(x)} \right]^2 dx$$

$$+ \frac{1}{n} \int_{-\infty}^{\infty} F_X(x)[1 - F_X(x)]dx - \frac{h}{n}\left[\frac{2(a^4+1)}{(a^2-1)^2}r_1 + r_2 \right] + o\left(h^8 + \frac{h}{n} \right).$$

2.3   Simulation Results

In this section, the results of the simulation study are presented to support the theoretical discussion. The generated random samples are from the standard normal distribution, normal distribution with mean 1 and variance 2, Laplace Distribution with mean 0 and scale parameter 1, and Laplace Distribution with mean 1 and scale parameter 2. The size of each sample is 50, with 1000 repetitions for each case.

Cross-validation is the method of choice for determining the bandwidths, and the kernel function used is the Gaussian Kernel function. Average Integrated Squared Error (ASE) as an estimator for MISE is calculated, and it is compared with the standard kernel distribution function estimator $\widehat{F}_h(x)$ and the proposed estimator $\tilde{F}_X(x)$ with several numbers of a's. The result can be seen in Table 2.1.

As can be seen, the proposed method gives good results, especially if the smaller a is used. However, at some point, the differences become smaller and smaller. That is why it is unnecessary to use too small a.

The next study done is drawing the graphs of the proposed method with several as ($a = 0.01$, $a = 0.25$, and $a = 3$), and comparing them with the graph of the true distribution function of the sample. The purpose of doing this is to see the effect of different as in the estimations. Generated sample with size 50 from the normal distribution with mean 1 and variance 2 is used. The method of choosing bandwidths and the kernel function are the same as before.

Table 2.1 AISE of standard and proposed method

Estimators	$N(0, 1)$	$N(1, 2)$	$Lap(0, 1)$	$Lap(1, 2)$
standard	0.06523	0.07502	0.09098	0.08511
$a = 0.01$	0.03106	0.034894	0.03043	0.04096
$a = 0.1$	0.03127	0.035002	0.03066	0.04149
$a = 0.25$	0.03199	0.0397	0.04353	0.04488
$a = 0.5$	0.04837	0.0469	0.04499	0.04902
$a = 0.75$	0.04917	0.04761	0.04940	0.04947
$a = 2$	0.05415	0.05017	0.06899	0.06760
$a = 3$	0.05745	0.05032	0.0695	0.06826

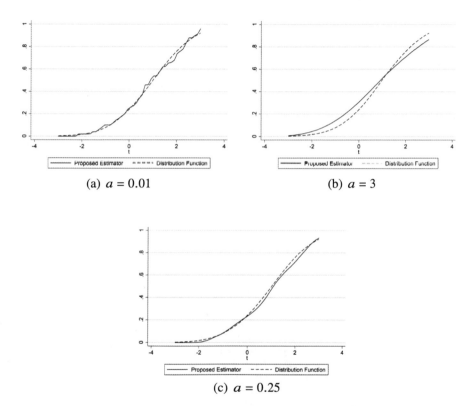

(a) $a = 0.01$ (b) $a = 3$

(c) $a = 0.25$

Fig. 2.1 Graphical comparison of the theoretical $N(1, 2)$ distribution vs $\tilde{F}_X$.

Even though Fig. 2.1a gives the most accurate estimation, but the graph is not as smooth as it should be and is similar to a step function. On the other hand, Fig. 2.1b gives the smoothest line, but it is the most inaccurate graph among all three.

In this simulation study, the best one is when $a = 0.25$ (Fig. 2.1c). It is smooth enough and the accuracy is not so different with Fig. 2.1a. It does not mean in general

$a = 0.25$ is the best one, but from this, it can be concluded that choosing a should be done wisely to get the desired result.

Even though in this book that $\tilde{F}_X \to F_n$ when $a \to 0$ is not proven mathematically, the similarity of Fig. 2.1a with a step function provides an intuition that when a is very close to 0, the behavior of the proposed estimator is becoming more similar to the empirical distribution function.

References

1. Serfling RJ (1980) Approximation theorems of mathematical statistics. Wiley, New York
2. Nadaraya EA (1964) Some new estimates for distribution functions. Theory Probab Appl 15:497–500
3. Winter BB (1973) Strong uniform consistency of integrals of density estimators. Can J Stat 1:247–253
4. Yamato H (1973) Uniform convergence of an estimator of a distribution function. Bull Math Stat 15:69–78
5. Yukich JE (1989) A note on limit theorems for pertubed empirical processes. Stoch Process Appl 33:163–173
6. Watson GS, Leadbetter MR (1964) Hazard analysis II. Sankhya A 26:101–106
7. Winter BB (1979) Convergence rate of perturbed empirical distribution functions. J Appl Probab 16:163–173
8. Degenhardt HJA (1993) Chung-Smirnov property for perturbed empirical distribution functions. Stat Probab Lett 16:97–101
9. Azzalini A (1981) A note on the estimation of a distribution function and quantiles by a kernel method. Biometrika 68:326–328
10. Reiss RD (1981) Nonparametric estimation of smooth distribution functions. Scand J Stat 8:116–119
11. Falk M (1983) Relative efficiency and deficiency of kernel type estimators of smooth distribution functions. Stat Neerl 37:73–83
12. Singh RS, Gasser T, Prasad B (1983) Nonparametric estimates of distribution functions. Commun Stat Theory Method 12:2095–2108
13. Hill PD (1985) Kernel estimation of a distribution function. Commun Stat Theory Method 14:605–620
14. Swanepoel JWH (1988) Mean integrated squared error properties and optimal kernels when estimating a distribution function. Commun Stat Theory Method 17:3785–3799
15. Shirahata S, Chu IS (1992) Integrated squared error of kernel type estimator of distribution function. Ann Inst Stat Math 44:579–591
16. Abdous B (1993) Note on the minimum mean integrated squared error of kernel estimates of a distribution function and its derivatives. Commun Stat Theory Method 22:603–609
17. Fauzi RR, Maesono Y (2017) Error reduction for kernel distribution function estimators. Bull Inform Cybern 49:53–66

Chapter 3
Kernel Quantile Estimation

Abstract From kernel distribution function, quantile estimators can be defined naturally. Using the kernel estimator of the p-th quantile of a distribution brings about an improvement in comparison to the sample quantile estimator. The improvement is revealed when applying the Edgeworth expansion of the kernel estimator. Using one more term beyond the normal approximation significantly improves the accuracy for small to moderate samples. The investigation is nonstandard since the influence function of the resulting L-statistic explicitly depends on the sample size. We obtain the expansion, justify its validity and demonstrate the numerical gains in using it.

Quantile estimation is an important problem in statistical inference. This is mostly due to the important practical applications of the method in the financial industry and risk assessment. The risk measure VaR (value at risk) is a quantile of the population distribution. It is known that the Lorenz curve [1], as an important indicator in econometrics for income distributions, also represents the transformation of quantiles. Further, estimating the quantile itself as accurately as possible is an essential tool for inference; it is important in its own right and will be dealt with in this paper.

3.1 Quantile Estimators

For a continuous random variable X with a cumulative distribution function $F(\cdot)$, density function $f(\cdot)$ and $E|X| < \infty$, the p-th quantile is defined as $Q(p) = \inf\{x : F(x) \geq p\}$. Let $X_1, X_2, \ldots, X_n$, be independently and identically distributed random variables with the distribution function $F(\cdot)$, and the simplest estimator of $Q(p)$ is the sample quantile estimator of the empirical distribution function

$$\widehat{Q}(p) = \inf\{x : F_n(x) \geq p\}$$

where $F_n(x) = \frac{1}{n} \sum_{i=1}^{n} I(X_i \leq x)$. It is a very popular estimator that is also typically implemented in statistical packages. Under mild conditions, it is asymptotically normal, and its asymptotic variance is $\sigma^2 = \frac{p(1-p)}{f^2(Q(p))}$ which happens to be large, par-

© The Author(s), under exclusive license to Springer Nature Singapore Pte Ltd. 2023 29
Rizky Reza Fauzi and Y. Maesono, *Statistical Inference Based on Kernel Distribution Function Estimators*, JSS Research Series in Statistics,
https://doi.org/10.1007/978-981-99-1862-1_3

ticularly, in the tails of the distribution where the density $f(x)$ has small values. In risk management, the tails are important regions of interest.

An alternative estimator is the kernel quantile estimator

$$\widehat{Q}_{p,h} = \frac{1}{h} \int_0^1 F_n^{-1}(x) K\left(\frac{x-p}{h}\right) dx, \tag{3.1}$$

where $F_n^{-1}(x)$ denotes the inverse of the empirical distribution function, and $K(\cdot)$ is a suitably chosen kernel. Of course, the bandwidth should be chosen appropriately. Besides the obvious requirement $h \to 0 (n \to \infty)$, additional requirements— in combination with requirements on the kernel—must be placed to ensure consistency, asymptotic normality, asymptotic bias elimination, and higher order accuracy of the estimator $\widehat{Q}_{p,h}$.

3.2 Properties of Quantile Estimators

Let us define order statistics $X_{1:n} \leq X_{2:n} \leq X_{n:n}$ based on the sample $X_1, X_2, \ldots, X_n$. Then the estimator $\widehat{Q}(p)$ is asymptotically equivalent to $X_{[np]+1:n}$ where $[a]$ denotes a maximum integer less than a. Further let us define order statistics $U_{1:n} \leq U_{2:n} \leq U_{n:n}$ based on a random sample $U_1, U_2, \ldots, U_n$ from the uniform distribution $U(0, 1)$. Then $X_{[np]+1:n}$ and $F^{-1}(U_{[np]+1:n})$ are asymptotically equivalent and we can discuss asymptotic expectation and variance of $\widehat{Q}(p)$. Since the distribution of $U_{[np]+1:n}$ is a beta distribution $Be([np] + 1, n - [np])$, we can obtain the asymptotic expectation and variance as follows. Expanding around p, we have

$$E[F^{-1}(U_{[np]+1:n})] = E\Big[F^{-1}(p) + (U_{[np]+1:n} - p)\{f(F^{-1}(p))\}^{-1}$$
$$-\frac{1}{2}(U_{[np]+1:n} - p)^2 \frac{f'(F^{-1}(p))}{f^3(F^{-1}(p))}\Big] + O(n^{-2})$$
$$= F^{-1}(p) + \frac{1}{f(F^{-1}(p))} E[U_{[np]+1:n} - p]$$
$$-\frac{1}{2}\frac{f'(F^{-1}(p))}{f^3(F^{-1}(p))} E[(U_{[np]+1:n} - p)^2] + O(n^{-2}).$$

Thus, using the moments of the beta distribution, we can show that

$$E[F^{-1}(U_{[np]+1:n})] = F^{-1}(p) + \frac{A}{f(F^{-1}(p))} - \frac{1}{2}\frac{f'(F^{-1}(p))}{f^3(F^{-1}(p))} B + O(n^{-2})$$

where

$$A = \frac{[np] + 1}{n + 1} - p,$$

$$B = \frac{([np] + 2)([np] + 1)}{(n + 2)(n + 1)} - 2p \left(\frac{[np] + 1}{n + 1} \right) + p^2.$$

Since $|[np] - np| \le 1$ and $F^{-1}(p) = Q(p)$, we have that $A = O(n^{-1})$ and $B = O(n^{-1})$. Then the bias of $\widehat{Q}(p)$ is $O(n^{-1})$. Similarly, we can obtain the variance of the estimator $\widehat{Q}(p)$

$$V(\widehat{Q}(p)) = \frac{p(1 - p)}{nf^2(F^{-1}(p))} + O(n^{-2}), \tag{3.2}$$

and then an asymptotic mean squared error with residual term $O(n^{-2})$ is given by

$$AMSE(\widehat{Q}(p)) = AMSE(F^{-1}(U_{[np]+1;n})) = \frac{p(1 - p)}{f^2(F^{-1}(p))n} = \frac{Q'(p)^2 p(1 - p)}{n}. \tag{3.3}$$

The estimator in (3.1) has been studied by many researchers in the past (see [2–6] and the references therein). Clearly, $\widehat{Q}_{p,h}$ is an L-estimator since it can be written as a weighted sum of the order statistics $X_{i:n}, i = 1, 2, \ldots, n$:

$$\widehat{Q}_{p,h} = \sum_{i=1}^{n} v_{i,n} X_{i:n}, \qquad v_{i,n} = \frac{1}{h} \int_{\frac{i-1}{n}}^{\frac{i}{n}} K \left(\frac{x - p}{h} \right) dx. \tag{3.4}$$

The difficulty in using standard asymptotic theory about L-statistics when analyzing the behavior of $\widehat{Q}_{p,h}$ is related to the fact that the weights $v_{i,n}$ in (3.4) depend explicitly on n in a peculiar way and tend to vanish asymptotically. Standard statements about asymptotic expansions of L-statistics (e.g., [1, 7]) are related to the case where the L-statistic could be written as $\int_0^1 F_n^{-1}(u) J(u) du$ with the *score function* $J(u)$ (which itself is an asymptotic limit) that does *not* involve n. The reason is that it is impossible to write its "score function" in such a way since it becomes a delta function in the limit. The treatment is technically involved and this has slowed down the research on the properties of $\widehat{Q}_{p,h}$ as an estimator of $Q(p)$. It is particularly interesting to derive higher order expansions for the asymptotic distribution of $\widehat{Q}_{p,h}$. Indeed, research by [2, 6] shows that the kernel quantile estimator is asymptotically equivalent to the empirical quantile estimator up to first order, hence any advantage would be revealed by involving at least one more term in the Edgeworth expansion and for moderate sample sizes. In this paper, we will derive such an expansion, will justify its validity, and will illustrate its advantages numerically. Since the empirical distribution $F_n(x)$ is a discrete random variable, we cannot prove the validity of its formal Edgeworth expansion.

A standard approach in the study of asymptotic properties of L-statistics is to first decompose them into an U-statistic plus a small-order remainder term and to apply asymptotic theory for the main term, that is, the U-statistic. When the $J(u)$ function depends on n, however, such a decomposition leads to an object that is similar to an asymptotic U-statistic [8] yet its defining function $h(.,.)$ explicitly depends on n and standard results about U-statistics can not be used directly. Progress in this direction has been achieved by [5, 6]. By applying Esseen's smoothing lemma ([9] Chap. XVI) and exploiting a nice decomposition by Friedrich for the resulting statistic, he has shown that a Berry-Esseen-type result of the form

$$P(\sqrt{n}|\widehat{Q}_{p,h} - Q(p)| \leq x\sigma_n) = 2\Phi(x) - 1 + O(n^{-r}) \tag{3.5}$$

holds whereby σ_n^2 can be given explicitly in terms of the kernel K and the derivative $Q'(p)$ of $Q(p)$. The rate r depends on the order m of the kernel. When $m = 2$ we obtain $O(n^{-1/3})$, which is improved to $O(n^{-5/13})$ for $m = 3$. Naturally, for no kernel order m is it possible to get the reminder order in (3.5) down to $O(n^{-1/2})$, not to mention $o(n^{-1/2})$ without further including the next term from an Edgeworth expansion. In [10] (Lemma 1), the authors have attempted to derive such an expansion in order to improve (3.5). However, the result as stated there is incorrect. Both the definitons of the expansion function $(G_n(x))$ and the order of the expansion $(o(n^{-1/3}))$ do not make sense and the argument used in the derivation is imprecise. Indeed, the authors rely on a technique about U-statistics used by [11] but in the latter paper the function $h(\cdot, \cdot)$ in the U-statistic's definition does *not* depend on n.

The purpose of this paper is to suggest a correct Edgeworth expansion for the kernel-based quantile estimator $\widehat{Q}_{p,h}$ up to order $o(n^{-1/2})$. The derivation is non-trivial. The specific requirement on the bandwidth h for the sake of eliminating the asymptotic bias triggers the need to include further contributions from the terms of order $O(n^{-1})$ of the appropriately resulting U-statistic. These contributions have to be taken into account in order to achieve the desired Edgeworth expansion with remainder of order $o(n^{-1/2})$. Although the general result is involved, the expansion can be simplified significantly in the most typical and practically relevant case of a symmetric compactly supported kernel $K(\cdot)$. Finally, we present numerical evidence about the improved accuracy of the Edgeworth expansion in comparison to the normal approximation for moderate sample sizes and some common distributions $F(x)$.

The proofs of our results require novel dedicated approaches. After an extensive search of the literature, we were indeed able to find one recent result (Lemma 2.1 of [12]) concerning the Edgeworth expansion of uniformly integrable transformations of independent identically distributed random variables in the form

$$\frac{1}{\sqrt{n}} \sum_{i=1}^{n} V_n(X_i) + \frac{1}{n^{3/2}} \sum_{1 \leq i < j \leq n} W_n(X_i, X_j), n \geq 2. \tag{3.6}$$

whereby the transformations $V_n(\cdot)$ and $W_n(\cdot, \cdot)$ are explicitly allowed to depend on n. For the sake of completeness, we quote the statement of lemma:

Let $W_n(x, y)$ be a symmetric function in its arguments. Assume also that:

(a) $EV_n(X_1) = 0$, $EV_n^2(X_1) = 1$, $|V_n(X_1)|^3$ are uniformly integrable and the distribution of $V_n(X_1)$ is non-lattice for all sufficiently large n.

(b) $E(W_n(X_1, X_2)|X_1) = 0, |W_n(X_1, X_2)|^{5/3}$ are uniformly integrable and $|W_n(X_i, X_j)| \le n^{3/2}$ for all $n \ge 2$ $i \ne j$.
Then, as $n \to \infty$,

$$\sup_x \left| P\left(\frac{1}{\sqrt{n}} \sum_{i=1}^{n} V_n(X_i) + \frac{1}{n^{3/2}} \sum_{1 \le i < j \le n} W_n(X_i, X_j) \le x \right) - F_n^{(1)}(x) \right| = o(n^{-1/2}),$$

$$(3.7)$$

where $F_n^{(1)}(x) = \Phi(x) - \frac{\Phi^{(3)}(x)}{6\sqrt{n}}(EV_n^3(X_1) + 3EV_n(X_1)V_n(X_2)W_n(X_1, X_2))$. Here, $\Phi(x)$ denotes the cdf of the standard normal distribution, $\phi(x)$ is its density, and $\Phi^{(3)}(x)$ is the third derivative.

However, as already pointed out, the above Lemma does not quite serve our purposes since we need one more term in the expansion of the U-statistic.

3.3 Asymptotic Properties of Quantile Estimators

For simplicity of the exposition, we will be using a compactly supported kernel $K(x)$ on $(-1, 1)$. We will say that the kernel is of order m if

$$K(x) \in L^2(-\infty, \infty), K^{(m)}(x) \in Lip(\alpha) \text{ for some } \alpha > 0,$$

$$\int_{-1}^{1} K(x)dx = 1, \int_{-1}^{1} x^i K(x)dx = 0, i = 1, 2, \ldots, m-1, \int_{-1}^{1} x^m K(x)dx \ne 0$$

For now, we do not require symmetry of the kernel.

Achieving a fine bias-variance trade-off for the kernel-based estimator is a delicate matter. It will be seen in the proof that the proper choice of bandwidth is given by

$$h = o(n^{-1/4}) \quad \text{and} \quad \lim_{n \to \infty} (n^{1/4}h)^{-k} n^{-\beta} = 0 \tag{3.8}$$

for any $\beta > 0$ and integer k. The typical bandwidth we choose to work with in the numerical implementation is $h = n^{-1/4}(\log n)^{-1}$.

Before formulating the main results of the paper, we briefly review the moment evaluations of the H-decomposition. For independent identically distributed random variables $X_1, \ldots, X_n$ and a function $v(x_1, \ldots, x_r)$ which is symmetric in its arguments with the property $E[v(X_1, \ldots, X_r)] = 0$, we define

$$\rho_1(x_1) = E[v(x_1, X_2, \ldots, X_r)],$$
$$\rho_2(x_1, x_2) = E[v(x_1, x_2, \ldots, X_r)] - \rho_1(x_1) - \rho_1(x_2), \ldots,$$

and

$$\rho_r(x_1, x_2, \ldots, x_r) = \nu(x_1, x_2, \ldots, x_r) - \sum_{k=1}^{r-1} \sum_{C_{r,k}} \rho_k(x_{i_1}, x_{i_2}, \ldots, x_{i_k})$$

where $\sum_{C_{r,k}}$ indicates that the summation is taken over all integers $i_1, \ldots, i_k$ satisfying $1 \le i_1 < \cdots < i_k \le r$. Then it holds

$$E[\rho_k(X_1, \ldots, X_k)|X_1, \ldots, X_{k-1}] = 0 \quad a.s. \tag{3.9}$$

and

$$\sum_{C_{n,r}} \nu(X_{i_1}, \ldots, X_{i_r}) = \sum_{k=1}^{r} \binom{n-k}{r-k} A_k \tag{3.10}$$

where

$$A_k = \sum_{C_{n,k}} \rho_k(X_{i_1}, \ldots, X_{i_k}). \tag{3.11}$$

Using Eq. (3.9) and the moment evaluations of martingales [13], we have the upper bounds of the absolute moments of A_k. For $q \ge 2$, if $E|\nu(X_1, \ldots, X_r)|^q < \infty$, there exists a positive constant C, which may depend on ν and F but not on n, such that

$$E|A_k|^q \le C n^{qk/2} E|\rho_k(X_{i_1}, \ldots, X_{i_k})|^q. \tag{3.12}$$

Since the empirical distribution $F_n(u)$ is a sum of independent identically distributed (i.i.d.) random variables, we have moment evaluations of $|F_n(u) - F(u)|$. For $q \ge 2$, we get

$$E|F_n(u) - F(u)|^q \le C n^{-q/2} F(u)(1 - F(u)) \tag{3.13}$$

where C is a constant. Hereafter, C will denote a generic constant that may change its meaning at different places in the text.

Using the inequalities (3.12) and (3.13), we will obtain an asymptotic representation of the standardized quantile estimator with residual $o_L(n^{-1/2})$

$$P(|o_L(n^{-1/2})| \ge \gamma_n n^{-1/2}) = o(n^{-1/2})$$

where $\gamma_n \to 0$ as $n \to \infty$.

Let $\{Y_i\}_{i=1,\ldots,n}$ be independent random variables uniformly distributed on $(0, 1)$ and define

$$\bar{Q}(p) = \frac{1}{h} \int_0^1 F^{-1}(x) K(\frac{x-p}{h}) dx,$$

$$\hat{I}_x(Y_1) = I(Y_1 \le p + hx) - (p + hx),$$

$$g_{1n}(Y_1) = -\int_{-1}^1 Q'(p + hx) K(x) \hat{I}_x(Y_1) dx,$$

$$\sigma_n^2 = Var(g_{1n}(Y_1)),$$

$$d_{1n} = \sigma_n^{-1} n^{-1/2}, \qquad d_{2n} = \sigma_n^{-1} n^{-3/2} h^{-1}, \qquad d_{3n} = \sigma_n^{-1} n^{-5/2} h^{-2},$$

$$g_{2n}(Y_1, Y_2) = -\int_{-1}^1 Q'(p + hx) K'(x) \hat{I}_x(Y_1) \hat{I}_x(Y_2) dx,$$

$$g_{3n}(Y_1, Y_2, Y_3) = -\int_{-1}^1 Q'(p + hx) K^{(2)}(x) \hat{I}_x(Y_1) \hat{I}_x(Y_2) \hat{I}_x(Y_3) dx,$$

$$\hat{g}_{1n}(Y_1) = -\frac{1}{2} \int_{-1}^1 Q'(p + hx) K^{(2)}(x) E[\hat{I}_x^2(Y_2)] \hat{I}_x(Y_1) dx,$$

$$A_{1n} = \sum_{i=1}^n g_{1n}(Y_i), \qquad A_{2n} = \sum_{C_{n,2}} g_{2n}(Y_i, Y_j), \qquad A_{3n} = \sum_{C_{n,3}} g_{3n}(Y_i, Y_j, Y_k)$$

and $\hat{A}_{1n} = \sum_{i=1}^n \hat{g}_{1n}(Y_i)$. Then we have the following lemma.

Lemma 3.1 *Assume* $\int [F(x)(1 - F(x))]^{1/5} dx < \infty$. *Let* $Q^{(m)}$ *be uniformly bounded in a neighborhood of* p $(0 < p < 1)$ *and* $f(Q(p)) > 0$. *Let* $K(x)$ *be a fourth-order kernel (i.e.,* $m = 4$) *and* $K^{(4)}(x) \in Lip(\alpha), \alpha > 0$. *Denote* $\delta = Q'(p)(\frac{1}{2} - p) + \frac{1}{2} Q^{(2)}(p) p(1 - p)$. *Further, choose* h *satisfying (3.8). Then we have*

$$\sigma_n^{-1} \sqrt{n}(\widehat{Q}_{p,h} - \bar{Q}(p))$$

$$= d_{1n} A_{1n} + d_{2n} A_{2n} + d_{3n} A_{3n} + d_{3n}(n-1)\hat{A}_{1n} + \frac{\delta}{\sigma \sqrt{n}} + o_L(n^{-1/2}).$$

The lemma shows that in order to obtain the Edgeworth expansion of $\sigma_n^{-1} \sqrt{n}(\widehat{Q}_{p,h} - \bar{Q}(p))$, we can concentrate first on the Edgeworth expansion of

$$d_{1n} A_{1n} + d_{2n} A_{2n} + d_{3n} A_{3n} + d_{3n}(n-1)\hat{A}_{1n}$$

with residual term $o_L(n^{-1/2})$. We will also prove the validity of the Edgeworth expansion.

Theorem 3.1 *Under the assumptions of Lemma 3.1, we have*

$$P(\sqrt{n}(\widehat{Q}_{p,h} - \bar{Q}(p)) \le x\sigma_n) = G_n(x - \frac{\delta}{\sigma \sqrt{n}}) + o(n^{-1/2}). \qquad (3.14)$$

Here:

$$G_n(x) = \Phi(x) - \phi(x)\left\{\frac{x^2 - 1}{6n^{1/2}\sigma_n^3}\left(e_{1n} + \frac{3e_{2n}}{h}\right)\right.$$

$$\left. + \frac{1}{nh^2}\left(\frac{x}{4\sigma_n^2}\{4e_{5n} + e_{6n}\} + \frac{x^3 - 3x}{6\sigma_n^4}\{3e_{2n} + e_{4n}\} + \frac{x^5 - 10x^3 + 15x}{8\sigma_n^6}e_{2n}^2\right)\right\}.$$

where

$$e_{1n} = E[g_{1n}^3(Y_1)], \qquad e_{2n} = E[g_{1n}(Y_1)g_{1n}(Y_2)g_{2n}(Y_1, Y_2)],$$
$$e_{3n} = E[g_{1n}(Y_2)g_{1n}(Y_3)g_{2n}(Y_1, Y_2)g_{2n}(Y_1, Y_3)],$$
$$e_{4n} = E[g_{1n}(Y_1)g_{1n}(Y_2)g_{1n}(Y_3)g_{3n}(Y_1, Y_2, Y_3)],$$
$$e_{5n} = E[g_{1n}(Y_1)\hat{g}_{1n}(Y_1)] \quad \text{and} \quad e_{6n} = E[g_{2n}^2(Y_1, Y_2)].$$

Expanding $G_n(x - \frac{\delta}{\sigma\sqrt{n}})$ around x and keeping the $O(\frac{1}{\sqrt{n}})$ terms only, we can also write (3.14) as follows:

$$P(\sqrt{n}(\widehat{Q}_{p,h} - \bar{Q}(p)) \le x\sigma_n) = \Phi(x) - \phi(x)\left\{\frac{x^2 - 1}{6n^{1/2}\sigma_n^3}\left(e_{1n} + \frac{3e_{2n}}{h}\right)\right.$$

$$\left. + \frac{1}{nh^2}\left(\frac{x}{4\sigma_n^2}\{4e_{5n} + e_{6n}\} + \frac{x^3 - 3x}{6\sigma_n^4}\{3e_{2n} + e_{4n}\} + \frac{x^5 - 10x^3 + 15x}{8\sigma_n^6}e_{2n}^2\right)\right\}$$

$$- \frac{\delta}{\sigma\sqrt{n}}\phi(x) + o(n^{-1/2})$$

The obtained Edgeworth expansion (3.14) is not for the ultimate quantity of interest $\sqrt{n}(\widehat{Q}_{p,h} - Q(p))$. Representing

$$\sqrt{n}(\widehat{Q}_{p,h} - Q(p)) = \sqrt{n}(\widehat{Q}_{p,h} - \bar{Q}(p)) + d_n$$

with $d_n = \sqrt{n}(\bar{Q}(p) - Q(p)) = \sqrt{n}(\frac{1}{h}\int_0^1 F^{-1}(x)K(\frac{x-p}{h})dx - Q(p))$ we need to make sure that the asymptotic order of the bias d_n is kept under control. Substituting $x - p = yh$ and applying Taylor expansion up to order m of $Q(x) = F^{-1}(x)$ around p, we get

$$d_n = O(\sqrt{n}h^m)\int_{-1}^1 K(y)y^m dy.$$

Hence

$$P(\sqrt{n}(\widehat{Q}_{p,h} - Q(p)) \leq x\sigma_n) - G_n\left(x - \frac{\delta}{\sigma\sqrt{n}}\right) \tag{3.15}$$

$$= P(\sqrt{n}(\widehat{Q}_{p,h} - \bar{Q}(p)) \leq \left(x - \frac{d_n}{\sigma_n}\right)\sigma_n) - G_n\left(x - \frac{\delta}{\sigma\sqrt{n}}\right)$$

$$= G_n\left(x - \frac{d_n}{\sigma_n} - \frac{\delta}{\sigma\sqrt{n}}\right) - G_n(x - \frac{\delta}{\sigma\sqrt{n}}) + o(n^{-1/2})$$

Expanding further G_n around the point $x - \frac{\delta}{\sigma\sqrt{n}}$, we finally get

$$P(\sqrt{n}(\widehat{Q}_{p,h} - Q(p)) \leq x\sigma_n) - G_n\left(x - \frac{\delta}{\sigma\sqrt{n}}\right) = \phi\left(x - \frac{\delta}{\sigma\sqrt{n}}\right)\left(-\frac{d_n}{\sigma_n}\right) + o(n^{-1/2}) \tag{3.16}$$

To keep the contribution of d_n under control, we require $d_n - o(n^{-1/2})$, that is, $\sqrt{n}h^m = o(n^{-1/2})$. We see now that the requirements $h = o(n^{-1/4})$, $m = 4$ that we put for another reason in Theorem 3.1, also guarantee that $d_n = o(n^{-1/2})$ holds. This is one more argument in favor of the choice $h = o(n^{-1/4})$ for the bandwidth. Moreover, the discussion in [4] (p. 411) also demonstrates from another point of view is that a reasonable choice of h should satisfy the order requirement $h = o(n^{-1/4})$. Their argument is that otherwise the dominant term in the MSE expansion of the kernel-quantile estimator may become worse than the one for the empirical quantile estimator, that is, $\frac{p(1-p)}{n}(Q'(p))^2$.

The above discussion together with Theorem 3.1, leads us to the following:

Theorem 3.2 *Under the assumptions of Theorem 3.2, we have the following Edgeworth expansion with remainder $o(n^{-1/2})$:*

$$P(\sqrt{n}(\widehat{Q}_{p,h} - Q(p)) \leq x\sigma_n) = G_n\left(x - \frac{\delta}{\sigma\sqrt{n}}\right) + o(n^{-1/2}) \tag{3.17}$$

Remark 3.1 Our expansion also leads to a Berry-Esseen bound

$$P(\sqrt{n}|\widehat{Q}_{p,h} - Q(p)| \leq x\sigma_n) = 2\Phi(x) - 1 + O(n^{-1/2})$$

This is an improvement on the result found by [5] whose bound for $m = 4$ only implies $P(\sqrt{n}|\widehat{Q}_{p,h} - Q(p)| \leq x\sigma_n) = 2\Phi(x) - 1 + O(n^{-7/17})$.

Remark 3.2 A more careful examination of the proof of Theorem 3.1 shows that for its statement to hold, one does not actually need the moment conditions $\int_{-1}^{1} x^i K(x)dx = 0$, $i = 1, 2, \ldots, m - 1$ on the kernel $K(x)$ to hold. However, these conditions are required in order to get the order of the bias right so that the ultimate expansion given in Theorem 3.2 could be obtained.

Remark 3.3 It is known that in terms of first-order performance *with respect to MSE* the kernel quantile estimator can only match the sample quantile [4]. The improvement with respect to the sample quantile can only show up in higher order terms of the MSE approximation (this phenomenon has been called *deficiency*). Much of the works of [2, 10] and others has been directed towards showing advantages of the kernel quantile estimator with respect to the deficiency criterion. In this paper, we approach the issue from another point of view by deriving the Edgeworth expansion for the non-lattice distribution of the kernel quantile estimator. This represents an alternative way of getting higher-order improvement in comparison to the sample quantile when estimating the quantile $Q(p)$. The Edgeworth expansion we suggest in this paper is more accurate than the normal approximation of the sample quantile and the normal approximation of the kernel quantile estimator.

Remark 3.4 Finally, we will comment on a crucial special case with important implications about the implementation of the Edgeworth expansion (3.17). We note that the general statement of Theorem 3.2 does not require *symmetry* (around zero) of the kernel $K(x)$. However, most of the kernels used in practical applications are usually symmetric [14] (p. 13). The symmetry assumption is also automatically made in Sheather and Marron [4] and in other influential papers on kernel quantile estimation. If we do assume that in addition the kernel *is* symmetric around zero (that is, $K(-x) = K(x)$ holds), then obviously $\int_{-1}^{1} K'(x)dx = 0$ holds. In that case, via Taylor expansion around p, we can see that the terms e_{2n} and e_{6n} in the expansion in Theorems 3.1 and 3.2 are of smaller order in comparison to the remaining terms and we get in this case the simpler expression:

$$P(\sqrt{n}(\widehat{Q}_{p,h} - Q(p)) \le x\sigma_n) = \Phi(x) - \phi(x)\left\{\frac{x^2 - 1}{6n^{1/2}\sigma_n^3}\left(e_{1n} + \frac{3e_{2n}}{h}\right)\right.$$

$$\left. + \frac{1}{nh^2}\left(\frac{xe_{5n}}{\sigma_n^2} + \frac{x^3 - 3x}{6\sigma_n^4}e_{4n}\right)\right\} - \frac{\delta}{\sigma\sqrt{n}}\phi(x) + o(n^{-1/2})$$

$$(3.18)$$

For higher order kernels like the ones we use in this paper, it is possible to make them also satisfy the additional condition $\int_{-1}^{1} K''(x)dx = 0$. If such kernel is chosen then expression (3.18) simplifies even further and becomes

$$P(\sqrt{n}(\widehat{Q}_{p,h} - Q(p)) \le x\sigma_n) = \Phi(x) - \phi(x)\frac{x^2 - 1}{6n^{1/2}\sigma_n^3}\left(e_{1n} + \frac{3e_{2n}}{h}\right)$$

$$- \frac{\delta}{\sigma\sqrt{n}}\phi(x) + o(n^{-1/2}).$$

$$(3.19)$$

Remark 3.5 We see the potentially most relevant practical applications of expansion (3.19) in evaluating power of tests of the type $H_0 : Q(p) = q_0$ against one-sided or two-sided alternatives. If the test is based on the statistic $\frac{\sqrt{n}(\widehat{Q}_{p,h} - q_0)}{\sigma_n}$ then by using the approximation (3.19) one should be able to get better power approximation for such a test in comparison to just using the normal approximation. Hereby, the approximation would be needed over the whole range of its values. Another application is in constructing more accurate confidence interval for the quantile $Q(p)$ when the sample size is small to moderate. For a given level α, instead of constructing it in a symmetric way as $\widehat{Q}_{p,h} \pm z_{\alpha/2}\sigma_n/\sqrt{n}$ one can improve the coverage accuracy by using $(\widehat{Q}_{p,h} + c_{1-\alpha/2}\sigma_n/\sqrt{n}, \widehat{Q}_{p,h} + c_{\alpha/2}\sigma_n/\sqrt{n})$ with the quantile values $c_{1-\alpha/2}$ and $c_{\alpha/2}$ obtained from the Edgeworth approximation.

3.4 Numerical Comparisons

Our goal in this section is to demonstrate numerically the effect of the improvement in the approximation of the distribution of the kernel quantile estimator when we move from the normal to the Edgeworth approximation. This effect could be seen for moderate sample sizes such as $n = 15, 30, 40, 50$ since for very large n the two approximations would become very close to each other. We could choose different kernels satisfying the requirements of Theorem 3.2 but the effect of the kernel is not that crucial. Here, we only present results obtained with the following symmetric fourth-order kernel first suggested by H. Müller:

$$K(x) = \frac{315}{512}(11x^8 - 36x^6 + 42x^4 - 20x^2 + 3)I(|x| \le 1).$$

The "asymptotically correct" bandwidth $h = n^{-1/4}(\log n)^{-1}$ turned out to be very well adjusted for samples of size $n \ge 50$. For smaller sample sizes (for which the $\log n$ is not small enough), it may be necessary to choose a smaller bandwidth, e.g., $0.1n^{-1/4}$, specifically when the value of p is near 0 or 1, to protect against a bias associated with edge effects.

The integral of the second derivative of this kernel is equal to zero and hence the approximation (3.19) can be applied when using it. The effect of the improvement from using Edgeworth expansion instead of the normal approximation depends on the particular distribution, on the sample size, and on the value of p. It is not to be expected to have a significant improvement when the underlying distribution is normal. Despite this, we also include the normal case in our simulations. More spectacular improvement could be expected for more skewed distributions, for smaller values of n (e.g., 15, 30), and for values of p closer to 0 or to 1. The subsections about the exponential and gamma distribution confirm this statement.

3.4.1 Estimation of Quantiles of the Standard Normal Distribution

In this case, we can derive easily:

$$Q'(p) = \frac{1}{\phi(\Phi^{-1}(p))}, \quad Q''(p) = \frac{\Phi^{-1}(p)}{\phi^2(\Phi^{-1}(p))}$$

We include the numerical values of "true" cumulative distribution function of the standardized random variable $\sqrt{n}(\widehat{Q}_{p,h} - Q(p))/\sigma_n$ and compare these values to the cumulative distribution of the standard normal and to the Edgeworth approximation as derived in Theorem 3.2. Table 3.1 gives the numerical values for two different scenarios: $n = 50$ and $n = 40$, both applied for the same value of $p = 0.1$. The "true" values of the cumulative distribution function were calculated on the basis of the empirical proportions for 2,000,000 repeated simulations. We found that at this number of simulations, the empirical ratios virtually do not change uniformly over the whole range $(-2.5, 2.5)$ of values of the argument up to the fourth decimal place. The values of e_{in} were calculated via averaging the numerical values of the resulting $g_{1n}(Y_1)$, $g_{2n}(Y_1, Y_2)$ values for 500,000 simulated independent uniform $(0,1)$ values Y_1, Y_2, whereas σ_n was calculated as the empirical standard deviation estimator for $\sqrt{Var(g_{1n}(Y_1))}$ using the 500,000 simulated Y_1 observations. Again, the 500,000

Table 3.1 Normal sample

v	Normal	Edgeworth	True value	Edgeworth	True
−2.4000	0.0082	0.0097	0.0091	0.0111	0.0091
−2.2000	0.0139	0.0155	0.0146	0.0174	0.0146
−2.0000	0.0228	0.0243	0.0227	0.0265	0.0227
−1.8000	0.0359	0.0370	0.0349	0.0392	0.0347
−1.6000	0.0548	0.0547	0.0519	0.0568	0.0515
−1.4000	0.0808	0.0787	0.0752	0.0802	0.0746
−1.2000	0.1151	0.1103	0.1061	0.1108	0.1052
−1.0000	0.1587	0.1505	0.1457	0.1496	0.1445
−0.8000	0.2119	0.1999	0.1944	0.1973	0.1936
−0.6000	0.2743	0.2586	0.2524	0.2542	0.2520
−0.4000	0.3446	0.3257	0.3192	0.3197	0.3194
−0.2000	0.4207	0.3998	0.3934	0.3927	0.3940
0.0000	0.5000	0.4783	0.4721	0.4708	0.4737
0.2000	0.5793	0.5583	0.5530	0.5512	0.5556
0.4000	0.6554	0.6366	0.6321	0.6306	0.6359
0.6000	0.7257	0.7101	0.7069	0.7056	0.7116
0.8000	0.7881	0.7762	0.7742	0.7735	0.7795
1.0000	0.8413	0.8332	0.8320	0.8322	0.8374
1.2000	0.8849	0.8802	0.8797	0.8807	0.8852
1.4000	0.9192	0.9172	0.9172	0.9187	0.9222
1.6000	0.9452	0.9451	0.9454	0.9472	0.9495
1.8000	0.9641	0.9651	0.9656	0.9674	0.9687
2.0000	0.9772	0.9788	0.9793	0.9810	0.9815
2.2000	0.9861	0.9877	0.9880	0.9896	0.9896
2.4000	0.9918	0.9933	0.9934	0.9947	0.9945

simulations were chosen on the basis of our experimentation showing that the resulting estimated parameter values for e_{in}, and for σ_n were virtually unchanged by a further increase in the number of simulations.

Table 3.1 shows the size of the improvement and demonstrates that except for a very tiny region of values in the lower tails of the distribution, the Edgeworth expansion approximated much better the true cumulative distribution function's values. In the small region of values in the lower tail where the opposite happened, the difference in the two approximations is negligibly small (it is in the third digit after the decimal point only, whereas, in the region where the Edgeworth expansion is better, the effect of the improvement is typically observed in the second digit).

3.4.2 Estimation of Quantiles of the Standard Exponential Distribution

The following calculations are easy to derive in this case:

$$Q(p) = -ln(1 - p), \, Q'(p) = \frac{1}{1 - p}, \, Q''(p) = \frac{1}{(1 - p)^2}.$$

Here, the improvement using the Edgeworth approximation is much more significant and can be felt for sample sizes as small as $n = 15$. Moreover, the improvement seems to be *virtually uniform* all over the whole range of values of the argument (except for a small region in the very left tail where, as is known, the Edgeworth expansion may become negative, in which case it should be set equal to zero). We include here the numerical values of "true" cumulative distribution function of the standardized random variable $\sqrt{n}(\widehat{Q}_{p,h} - Q(p))/\sigma_n$ and compare these values to the cumulative distribution of the standard normal and to the Edgeworth in Table 3.2. The numerical values are given for two different scenarios: $n = 30$ and $n = 15$, both applied for the same value of $p = 0.9$. Again, the "true" values of the cumulative distribution was calculated on the basis of the empirical proportions for 2,000,000 repeated simulations. The values of e_{in} were calculated via averaging the numerical values of the resulting $g_{1n}(Y_1)$, $g_{2n}(Y_1, Y_2)$ values for 500,000 simulated independent uniform (0,1) pairs of values Y_1, Y_2, whereas σ_n was calculated as the empirical standard deviation estimator for $\sqrt{Var(g_n(Y_1))}$ using the 500,000 simulated Y_1 observations.

Table 3.2 shows that the effect of the improvement from using the Edgeworth expansion is virtually over the whole range of values of the argument.

3.4.3 Estimation of Quantiles of the Gamma Distribution

There is a very reliable algorithmDM for calculating the inverse of the incomplete Gamma function. This alleviates a lot the testing of our approximation for any Gamma

Table 3.2 Exponential sample

v	Normal	Edgeworth	True value	Edgeworth	True
−2.4000	0.0082	−0.0052	0.0004	−0.0099	0.0000
−2.2000	0.0139	−0.0035	0.0015	−0.0095	0.0003
−2.0000	0.0228	0.0017	0.0045	−0.0055	0.0020
−1.8000	0.0359	0.0124	0.0117	0.0042	0.0078
−1.6000	0.0548	0.0307	0.0256	0.0223	0.0218
−1.4000	0.0808	0.0590	0.0494	0.0513	0.0483
−1.2000	0.1151	0.0989	0.0854	0.0930	0.0905
−1.0000	0.1587	0.1513	0.1351	0.1482	0.1487
−0.8000	0.2119	0.2155	0.1976	0.2161	0.2212
−0.6000	0.2743	0.2896	0.2712	0.2941	0.3032
−0.4000	0.3446	0.3704	0.3527	0.3783	0.3896
−0.2000	0.4207	0.4537	0.4376	0.4641	0.4761
0.0000	0.5000	0.5356	0.5220	0.5468	0.5582
0.2000	0.5793	0.6123	0.6019	0.6226	0.6338
0.4000	0.6554	0.6812	0.6747	0.6892	0.7006
0.6000	0.7257	0.7411	0.7390	0.7456	0.7585
0.8000	0.7881	0.7918	0.7944	0.7924	0.8074
1.0000	0.8413	0.8340	0.8408	0.8309	0.8479
1.2000	0.8849	0.8688	0.8784	0.8629	0.8810
1.4000	0.9192	0.8975	0.9083	0.8898	0.9073
1.6000	0.9452	0.9211	0.9319	0.9127	0.9285
1.8000	0.9641	0.9405	0.9500	0.9323	0.9452
2.0000	0.9772	0.9562	0.9636	0.9490	0.9581
2.2000	0.9861	0.9687	0.9737	0.9627	0.9680
2.4000	0.9918	0.9784	0.9812	0.9738	0.9756

distribution. For illustrative purposes, we show some results for the chi-squared distribution with 4 degree of freedom. This is a particular Gamma distribution with the density $f(x) = \frac{1}{4}xe^{-x/2}, x > 0$. After changing variables, we see that for this distribution, $Q(p)$ is a solution of the equation:

$$p = \int_0^{Q(p)/2} ye^{-y}dy, \tag{3.20}$$

which means that $Q(p)$ can be expressed via the inverse of an incomplete gamma function. The derivatives of $Q(p)$ are then easily obtained by using the $Q(p)$ value. Indeed, the following calculations follow easily by applying integration by parts in (3.20) and differentiating both sides of the equality with respect to p :

Table 3.3 Chi-square sample

v	Normal	Edgeworth	True	Edgeworth	True	Edgeworth	True
−2.4000	0.0082	0.0007	0.0026	−0.0005	0.0009	−0.0042	0.0004
−2.2000	0.0139	0.0042	0.0059	0.0026	0.0027	−0.0002	0.0016
−2.0000	0.0228	0.0111	0.0124	0.0092	0.0069	0.0036	0.0053
−1.8000	0.0359	0.0231	0.0238	0.0209	0.0154	0.0147	0.0138
−1.6000	0.0548	0.0419	0.0423	0.0397	0.0306	0.0334	0.0304
−1.4000	0.0808	0.0695	0.0698	0.0675	0.0553	0.0620	0.0589
−1.2000	0.1151	0.1074	0.1079	0.1059	0.0915	0.1021	0.1008
−1.0000	0.1587	0.1563	0.1575	0.1557	0.1406	0.1544	0.1576
−0.8000	0.2119	0.2162	0.2185	0.2166	0.2022	0.2185	0.2271
−0.6000	0.2743	0.2857	0.2888	0.2871	0.2745	0.2924	0.3061
−0.4000	0.3446	0.3622	0.3658	0.3646	0.3545	0.3729	0.3903
−0.2000	0.4207	0.4427	0.4470	0.4457	0.4385	0.4561	0.4750
0.0000	0.5000	0.5235	0.5281	0.5268	0.5226	0.5379	0.5571
0.2000	0.5793	0.6012	0.6061	0.6043	0.6025	0.6147	0.6332
0.4000	0.6554	0.6731	0.6782	0.6755	0.6762	0.6838	0.7010
0.6000	0.7257	0.7372	0.7427	0.7386	0.7417	0.7439	0.7602
0.8000	0.7881	0.7925	0.7986	0.7929	0.7981	0.7948	0.8101
1.0000	0.8413	0.8390	0.8456	0.8384	0.8447	0.8371	0.8513
1.2000	0.8849	0.8772	0.8836	0.8757	0.8827	0.8719	0.8848
1.4000	0.9192	0.9080	0.9139	0.9060	0.9129	0.9005	0.9118
1.6000	0.9452	0.9323	0.9375	0.9301	0.9364	0.9238	0.9329
1.8000	0.9641	0.9512	0.9555	0.9490	0.9541	0.9428	0.9493
2.0000	0.9772	0.9656	0.9687	0.9637	0.9673	0.9581	0.9619
2.2000	0.9861	0.9764	0.9782	0.9748	0.9770	0.9702	0.9716
2.4000	0.9918	0.9843	0.9852	0.9831	0.9840	0.9794	0.9790

$$Q'(p) = \frac{4 + 2Q(p)}{(1-p)Q(p)}, \quad Q''(p) = \frac{2Q(p)(2 + Q(p)) - 4Q'(p)(1-p)}{(1-p)^2 Q(p)^2}$$

In this case, the improvement due to the Edgeworth approximation is significant and can be felt for sample sizes as small as $n = 15$. Again, the improvement seems to be virtually uniform all over the whole range of values of the argument (except for a small region in the very left tail where the Edgeworth expansion may become negative, in which case it should be equal to zero). The numerical values of "true" cumulative distribution function of the standardized random variable $\sqrt{n}(\widehat{Q}_{p,h} - Q(p))/\sigma_n$ are compared to the cumulative distribution of the standard normal and to the Edgeworth in Table 3.3. The same number of simulations has been applied as in the previous examples. In order to demonstrate the increasing accuracy of both the Edgeworth and the normal approximation with increasing sample size, as well as the uniform improvement offered by the Edgeworth expansion for large n such as $n = 50$, the numerical values are given for three different scenarios: $n = 50$, $n = 30$, and $n = 15$, with $p = 0.9$. Again, it can be seen that the improvement is virtually over the whole range of values of the argument and that the gains, especially for $n = 15$, are impressive.

References

1. Lorenz M (1905) Methods of measuring concentration of wealth. J Am Stat Assoc 9:209–219
2. Falk M (1984) Relative deficiency of kernel type estimators of quantiles. Ann Stat 12(1):261–268
3. Falk M (1985) Asymptotic normality of the kernel quantile estimator. Ann Stat 13(1):428–433
4. Sheather S, Marron J (1990) Kernel quantile estimators. J Am Stat Assoc 85(410):410–416
5. Xiang X (1995) Deficiency of the sample quantile estimator with respect to kernel quantile estimators for censored data. Ann Stat 23(3):836–854
6. Xiang X (1995) A Berry-Esseen theorem for the kernel quantile estimator with application to studying the deficiency of qunatile estimators. Ann Inst Stat Math 47(2):237–251
7. Helmers R (1982) Edgeworth expansions for linear combinations of order statistics. Mathematical Centre Tracts (105). Amsterdam
8. Lai T, Wang J (1993) Edgeworth expansions for symmetric statistics with applications to bootstrap methods. Stat Sin 3:517–542
9. Feller W (1971) An introduction to probability and its applications, vol 2. Wiley, New York
10. Xiang X, Vos P (1997) Quantile estimators and covering probabilities. J Nonparametric Stat 7:349–363
11. Bickel P, Götze F, van Zwet W (1986) The edgeworth expansion for U-statistics of degree two. Ann Stat 14:1463–1484
12. Jing BY, Wang Q (2003) Edgeworth expansion for U-statistics under minimal conditions. Ann Stat 31(4):1376–1391
13. Dharmadhikari SW, Fabian V, Jogdeo K (1968) Bounds on the moments of martingales. Ann Math Stat 39:1719–1723
14. Silverman B (1986) Density estimation for statistics and data analysis. Chapman and Hall, London

Chapter 4
Mean Residual Life Estimator

Abstract Mean residual function is a function that represents the average remaining time before failure. Though not as severe as the boundary problems in kernel density estimation, eliminating the boundary bias problems that occur in the naive kernel estimator of the mean residual life function is needed. Here, the property of bijective transformation is once again utilized to define two boundary-free kernel-based mean residual life function estimators. Furthermore, the proposed methods preserve the mean value property, which cannot be done by the naive kernel estimator. Some simulation results showing the estimators' performances and real data analysis will be presented in the last part of this article.

Statistical inference for remaining lifetimes would be intuitively more appealing than the popular hazard rate function, since its interpretation as "the risk of immediate failure" can be difficult to grasp. A function called the mean residual life (or mean excess loss) which represents "the average remaining time before failure" is easier to understand. The mean residual life (or MRL for short) function is of interest in many fields related to time and finance, such as biomedical theory, survival analysis, and actuarial science.

Let $X_1, X_2, ..., X_n$ be independently and identically distributed absolutely continuous random variables supported on an interval $\Omega \subset \mathbb{R}$, where $\inf \Omega = a_1$, $\sup \Omega = a_2$, and $-\infty \leq a_1 < a_2 \leq \infty$. Also, let f_X be the density function, F_X be the cumulative distribution function, $S_X(t) = \Pr(X > t)$ be the survival function, and $\mathbb{S}_X(t) = \int_t^\infty S_X(x)\mathrm{d}x$ be the cumulative survival function, of X_i's. Then

$$m_X(t) = E(X - t | X > t), \quad t \in \Omega,$$

is the definition of the mean residual life function, or can be written as

$$m_X(t) = \frac{\mathbb{S}_X(t)}{S_X(t)}.$$

Detailed properties of the MRL function have been discussed in [1], including its usage for ordering and classifying distributions in [2], also a method to determine

© The Author(s), under exclusive license to Springer Nature Singapore Pte Ltd. 2023 45
Rizky Reza Fauzi and Y. Maesono, *Statistical Inference Based on Kernel Distribution Function Estimators*, JSS Research Series in Statistics,
https://doi.org/10.1007/978-981-99-1862-1_4

distribution via an inversion formula of m_X in [3]. The problem of characterization of the cumulative distribution function can be solved as well through the relationship between the MRL function and the hazard rate function [4]. The MRL functions of finite mixtures and order statistics have been studied in [5], and its further properties derived from proportional odds relevation was recently discussed in [6].

Some properties of the MRL concept related to operational research and reliability theory in engineering are interesting topics, such as the properties of the MRL function for associated orderings [7], its usefulness for maintenance decision-making [8], its utilization on a parallel system [9], records [10], a k-out-of-n:G system [11], a $(n - k + 1)$-out-of-n system [12], and the reliability shock models [13]. The MRL function is also subjected to Marshall-Olkin-type shocks [14], coherent systems [15], degrading systems [16], and rail wagon bearings [17]. The mean residual life function can also be used for estimating joint pricing in queueing system [18] and analyzing the mining truck tire's reliability performance [19].

The natural estimator of the MRL function is the empirical one, defined as

$$m_n(t) = \frac{\mathbb{S}_n(t)}{S_n(t)} = \frac{\sum_{i=1}^{n}(X_i - t)I(X_i > t)}{\sum_{i=1}^{n} I(X_i > t)}, \quad t \in \Omega, \tag{4.1}$$

where $I(A)$ is the usual indicator function on set A. Even though it has several good attributes such as unbiasedness and consistency [20], the empirical MRL function is just a rough estimate of m_X and lack of smoothness. Estimating is also impossible for large t because $S_n(t) = 0$ for $t > \max\{x_i\}$. Though defining $m_n(t) = 0$ for such a case is possible, it is a major disadvantage as analyzing the behavior of the MRL function for large t is of an interest.

Various parametric models have been discussed in literatures, for example, the transformed parametric MRL models [21], the upside-down bathtub-shaped MRL model [22], the MRL order of convolutions of heterogeneous exponential random variables [23], the proportional MRL model [24], also the MRL models with time-dependent coefficients [25]. Some lifetime-model distributions with interesting MRL function properties have been studied as well, such as the Harmonic Mixture Fréchet Distribution [26], the Quasi-Suja Distribution [27], the polynomial-exponential distribution [28], the Alpha Power Exponentiated Weibull-Pareto Distribution [29], also the Odd Modified Burr-III Exponential Distribution [30].

Some nonparametric estimators of m_X which are related to the empirical one have been discussed in a fair amount of literatures. For example, the numerator in m_n can be estimated by a recursive kernel estimate and left the empirical survival function unchanged [31], while Hille's Theorem is used to smooth out both the numerator and denominator in m_n [32]. The MRL function estimator and its confidence interval are also successfully constructed using modified empirical likelihood [33].

The other maneuver that can be used for estimating the MRL function nonparametrically is the kernel method. Let K be a symmetric continuous nonnegative kernel function with $\int_{-\infty}^{\infty} K(x)dx = 1$, and $h > 0$ be a bandwidth satisfying $h \to 0$ and $nh \to \infty$ when $n \to \infty$. From this, three other functions derived from K, which are

$$W(x) = \int_{-\infty}^{x} K(z)\mathrm{d}z, \quad V(x) = \int_{x}^{\infty} K(z)\mathrm{d}z, \quad \text{and} \quad \mathbb{V}(x) = \int_{x}^{\infty} V(z)\mathrm{d}z.$$

Hence, the naive kernel MRL function estimator can be defined as

$$\widehat{m}_X(t) = \frac{\widehat{\mathbb{S}}_X(t)}{\widehat{S}_X(t)} = \frac{h \sum_{i=1}^{n} \mathbb{V}\left(\frac{t-X_i}{h}\right)}{\sum_{i=1}^{n} V\left(\frac{t-X_i}{h}\right)}, \quad t \in \Omega, \tag{4.2}$$

with its asymptotic properties of the naive kernel MRL function estimator have been discussed in detail in [34].

However, as m_X is usually used for time- or financial-related data, which are on nonnegative real line or bounded interval, the naive kernel MRL function estimator suffers the so-called boundary bias problem. In the case of $f_X(a_1) = 0$ and $f_X(a_2) = 0$, the boundary effects of $\widehat{m}_X$ when $t \to a_1$ or $t \to a_2$ are not as bad as in the kernel density estimator, but the problems still occur, as it can be proven from the asymptotic formula that $Bias[\widehat{m}_X(t)] = O(h)$ when t is in the boundary [34]. It is due to $S_X(a_1) = 1 - S_X(a_2) = 1$ that the $Bias[\widehat{\mathbb{S}}_X(a_1)]$ and $Bias[\widehat{\mathbb{S}}_X(a_2)]$ can never be 0, which implies $\widehat{\mathbb{S}}_X$ causes the boundary problems for $\widehat{m}_X$. Moreover, in the case of $f_X(a_1) > 0$ and $f_X(a_2) > 0$ (e.g., uniform distribution), not only $\widehat{\mathbb{S}}_X$ but also $\widehat{S}_X$ adds its share to the boundary problems. In this extreme situation, $Bias[\widehat{m}_X(t)] = O(1)$ in the boundary regions.

To make things worse, the naive kernel MRL function estimator does not preserve one of the most important properties of the MRL function, which is $m_X(a_1) + a_1 = E(X)$ (mean value property). It is reasonable if $\widehat{m}_X(a_1) + a_1 \approx \bar{X}$ to be expected. However, $\widehat{S}_X(a_1)$ is less than 1 and $\widehat{\mathbb{S}}_X(a_1)$ is smaller than the average value of X_i's, due to the weight they still put on the outside of Ω. Hence, it explains why $\lim_{t \to a_1^+} \widehat{m}_X(t) = \bar{X} + O_p(1)$ in extreme boundary problem case.

In this book, another idea suggested by [35] was used to remove the boundary bias problem, which is utilizing transformations that map Ω to $\mathbb{R}$ bijectively. In this situation, there is no boundary effect at all, as any weight will not be put outside the support. Hence, instead of using $X_1, X_2, ..., X_n$, the kernel method on the transformed $Y_1, Y_2, ..., Y_n$, where $Y_i = g^{-1}(X_i)$ and $g : \mathbb{R} \to \Omega$ is a bijective function, will be applied. However, even though the idea is easy to understand, just substituting t with $g^{-1}(t)$ and X_i with Y_i in the formula of $\widehat{m}_X$ are not allowed, due to avoiding nonintegrability. Modifying the naive kernel MRL function estimator before substituting $g^{-1}(t)$ and Y_i in order to preserve the integrability and to ensure that the new formulas are good estimators is needed.

Before moving on to the main focus, there are some imposed conditions:

C1. the kernel function K is a continuous nonnegative function, symmetric at the origin, with $\int_{-\infty}^{\infty} K(x)\mathrm{d}x = 1$,
C2. the bandwidth $h > 0$ satisfies $h \to 0$ and $nh \to \infty$ when $n \to \infty$,
C3. the function $g : \mathbb{R} \to \Omega$ is continuous and strictly increasing,
C4. the density f_X and the function g are continuously differentiable at least twice,

C5. the integrals $\int_{-\infty}^{\infty} g'(ux)K(x)dx$ and $\int_{-\infty}^{\infty} g'(ux)V(x)dx$ are finite for all u in an ε-neighborhood of the origin,

C6. the expectations $E(X)$, $E(X^2)$, and $E(X^3)$ exist.

The first and the second conditions are standard assumptions for kernel methods, and C3 is needed for the bijectivity and the simplicity of the transformation. Since some serial expansions will be used, C4 is important to ensure the validity of the proofs. The last two conditions are necessary to make sure the possibility to derive the bias and the variance formulas. In order to calculate the variances, a new function

$$\bar{\mathbb{S}}_X(t) = \int_t^\infty \mathbb{S}_X(x)dx$$

is defined for simpler notation. Also, some lemmas are needed in the discussion though sometimes are not stated explicitly.

Lemma 4.1 *Under condition C1, the following equations hold*

$$\int_{-\infty}^\infty V(x)K(x)dx = \frac{1}{2},$$

$$\int_{-\infty}^\infty xV(x)K(x)dx = -\frac{1}{2}\int_{-\infty}^\infty V(x)W(x)dx,$$

$$\int_{-\infty}^\infty \mathbb{V}(x)K(x)dx = \int_{-\infty}^\infty V(x)W(x)dx.$$

Lemma 4.2 *Let $f_Y(t)$ and $S_Y(t)$ be the probability density function and the survival function of $Y = g^{-1}(X)$, also let $a(t) = \int_t^\infty g'(y)S_Y(y)dy$ and $A(t) = \int_t^\infty g'(y)a(y)dy$. Then, under the condition C6, for $t \in \Omega$,*

$$f_Y(g^{-1}(t)) = g'(g^{-1}(t))f_X(t), \tag{4.3}$$

$$S_Y(g^{-1}(t)) = S_X(t), \tag{4.4}$$

$$a(g^{-1}(t)) = \mathbb{S}_X(t), \tag{4.5}$$

$$A(g^{-1}(t)) = \bar{\mathbb{S}}_X(t). \tag{4.6}$$

Remark 4.1 The ideas to construct the proposed estimators actually came from Lemma 4.2. The estimators of S_X and $\mathbb{S}_X$ were constructed using the relationships stated in Eq. (4.4) and (4.5), respectively. This lemma is called the change-of-variable properties.

Lemma 4.3 *Let $a(t) = \int_t^\infty g'(y)S_Y(y)dy$ and*

$$\hat{a}(t) = \int_t^\infty g'(y)\hat{S}_Y(y)dy = \frac{1}{n}\sum_{i=1}^n \int_t^\infty g'(y)V\left(\frac{y - Y_i}{h}\right)dy$$

be the naive kernel estimator of a. If $B \subset \mathbb{R}$ is an interval where both $\widehat{a}$ and a are bounded, then $\sup_{t \in B} |\widehat{a}(t) - a(t)| \to_{a.s.} 0$.

Proof Since $\widehat{a}$ and a are both bounded, non-increasing, and continuous on B, then for any $\varepsilon > 0$, k number of points on B can be found such that

$$-\infty \leq \inf B = t_1 < t_2 < ... < t_k = \sup B \leq \infty,$$

and $a(t_j) - a(t_{j+1}) \leq \varepsilon/2$, $j = 1, 2, ..., k - 1$. For any $t \in B$, it is clear that there exists j such that $t_j \leq t < t_{j+1}$. For that particular j, then

$$\widehat{a}(t_j) \geq \widehat{a}(t) \geq \widehat{a}(t_{j+1}) \quad \text{and} \quad a(t_j) \geq a(t) \geq a(t_{j+1}),$$

which result in

$$\widehat{a}(t_{j+1}) - a(t_{j+1}) - \frac{\varepsilon}{2} \leq \widehat{a}(t) - a(t) \leq \widehat{a}(t_j) - a(t_j) + \frac{\varepsilon}{2}.$$

Therefore,

$$\sup_{t \in B} |\widehat{a}(t) - a(t)| \leq \sup_j |\widehat{a}(t_j) - a(t_j)| + \varepsilon.$$

Now, because $\widehat{a}(t)$ is a naive kernel estimator, it is clear that for fix t_0, $\widehat{a}(t_0)$ converges almost surely to $a(t_0)$. Thus, $|\widehat{a}(t_0) - a(t_0)| \to_{a.s.} 0$. Hence, for any $\varepsilon > 0$, almost surely $\sup_{t \in B} |\widehat{a}(t) - a(t)| \leq \varepsilon$ when $n \to \infty$, which concludes the proof. $\qquad \square$

4.1 Estimators of the Survival Function and the Cumulative Survival Function

Before jumping into the estimation of the mean residual life function, firstly, the estimations of its components will be discussed, which are the survival function S_X and the cumulative survival function $\mathbb{S}_X$. In this section, one survival function estimator and two cumulative survival function estimators using the idea of transformation will be proposed.

By generalizing the usage of probit transformation to eliminate the boundary bias problems in the kernel density estimation for data on the unit interval [36], utilizing any function g that satisfies conditions C3–C5 leads to

$$\widetilde{f}_X(t) = \frac{1}{nhg'(g^{-1}(t))} \sum_{i=1}^{n} K\left(\frac{g^{-1}(t) - g^{-1}(X_i)}{h}\right),$$

which is the boundary-free kernel density estimator [37]. Then, by doing simple substitution technique on $\int_t^{a_2} \widetilde{f}_X(x) dx$, the proposed survival function estimator is

$$\widetilde{S}_X(t) = \frac{1}{n}\sum_{i=1}^{n} V\left(\frac{g^{-1}(t) - g^{-1}(X_i)}{h}\right), \quad t \in \Omega. \tag{4.7}$$

By the same approach, the first cumulative survival function estimator is

$$\widetilde{\mathbb{S}}_{X,1}(t) = \frac{1}{n}\sum_{i=1}^{n} \mathbb{V}_{1,h}(g^{-1}(t), g^{-1}(X_i)), \quad t \in \Omega, \tag{4.8}$$

where

$$\mathbb{V}_{1,h}(x, y) = \int_{x}^{\infty} g'(z) V\left(\frac{z - y}{h}\right) dz. \tag{4.9}$$

Their biases and variances are given in the following theorem.

Theorem 4.1 *Under conditions C1 until C6, the biases and the variances of $\widetilde{S}_X(t)$ and $\widetilde{\mathbb{S}}_{X,1}(t)$ are*

$$Bias[\widetilde{S}_X(t)] = -\frac{h^2}{2}b_1(t)\int_{-\infty}^{\infty} y^2 K(y)dy + o(h^2), \tag{4.10}$$

$$Var[\widetilde{S}_X(t)] = \frac{1}{n}S_X(t)F_X(t) - \frac{h}{n}g'(g^{-1}(t))f_X(t)\int_{-\infty}^{\infty} V(y)W(y)dy + o\left(\frac{h}{n}\right), \tag{4.11}$$

and

$$Bias[\widetilde{\mathbb{S}}_{X,1}(t)] = \frac{h^2}{2}b_2(t)\int_{-\infty}^{\infty} y^2 K(y)dy + o(h^2), \tag{4.12}$$

$$Var[\widetilde{\mathbb{S}}_{X,1}(t)] = \frac{1}{n}[2\bar{\mathbb{S}}_X(t) - \mathbb{S}_X^2(t)] + o\left(\frac{h}{n}\right), \tag{4.13}$$

where

$$b_1(t) = g''(g^{-1}(t))f_X(t) + [g'(g^{-1}(t))]^2 f_X'(t), \tag{4.14}$$

$$b_2(t) = [g'(g^{-1}(t))]^2 f_X(t) + \int_{g^{-1}(t)}^{\infty} g''(x)g'(x)f_X(g(x))dx. \tag{4.15}$$

Furthermore, the covariance of them is

$$Cov[\widetilde{\mathbb{S}}_{X,1}(t), \widetilde{S}_X(t)] = \frac{1}{n}S_X(t)F_X(t) + o\left(\frac{h}{n}\right). \tag{4.16}$$

Proof The usual reasoning of i.i.d. random variables and the transformation property of expectation result in

$$E[\widetilde{S}_X(t)] = S_X(t) - \frac{h^2}{2}b_1(t)\int_{-\infty}^{\infty} u^2 K(u)du + o(h^2),$$

and this gives the $Bias[\widetilde{S}_X(t)]$. For the variance of $\widetilde{S}_X(t)$, first calculating

$$E\left[V^2\left(\frac{g^{-1}(t)-g^{-1}(X_1)}{h}\right)\right] = 2\int_{-\infty}^{\infty} S_Y(g^{-1}(t)-hu)V(u)K(u)du$$

$$= S_X(t) - hg'(g^{-1}(t))f_X(t)\int_{-\infty}^{\infty} V(u)W(u)du + o(h)$$

will lead to the desired formula.

For the calculation of $Bias[\widetilde{\mathbb{S}}_{X,1}(t)]$, note that

$$\mathbb{V}_{1,h}(g^{-1}(t), y) = \int_t^{a_2} V\left(\frac{g^{-1}(z)-y}{h}\right)dz,$$

and by assuming that the change of the order of the integral signs is allowed, hence

$$E[\widetilde{\mathbb{S}}_{X,1}(t)] = \int_t^{a_2} E\left[V\left(\frac{g^{-1}(z)-Y}{h}\right)\right]dz$$

$$= \mathbb{S}_X(t) - \frac{h^2}{2}\int_t^{a_2} b_1(z)dz\int_{-\infty}^{\infty} y^2 K(y)dy + o(h^2).$$

It is easy to see that $b_2(t) = -\int_t^{a_2} b_1(z)dz$, and the formula of $Bias[\widetilde{\mathbb{S}}_{X,1}(t)]$ is done. Before calculating $Var[\widetilde{\mathbb{S}}_{X,1}(t)]$, first, note that

$$\frac{d}{dy}\mathbb{V}_{1,h}(x, y) = \int_{\frac{x-y}{h}}^{\infty} g'(y+hz)K(z)dz = g'(y)V\left(\frac{x-y}{h}\right) + o(h)$$

and

$$\mathbb{V}_{1,h}(x, y) = h\int_{\frac{x-y}{h}}^{\infty} g'(y+hz)V(z)dz = hg'(y)\mathbb{V}\left(\frac{x-y}{h}\right) + o(h).$$

Now,

$$E[\mathbb{V}_{1,h}^2(g^{-1}(t), g^{-1}(X_1))] = 2\int_{-\infty}^{\infty}\left[g'(y)V^2\left(\frac{g^{-1}(t)-y}{h}\right)\right.$$

$$\left. +\frac{1}{h}\mathbb{V}_{1,h}(g^{-1}(t), y)K\left(\frac{g^{-1}(t)-y}{h}\right)\right]a(y)dy + o(h).$$

Conducting integration by parts once again for the first term, then

$$2 \int_{-\infty}^{\infty} g'(y) V^2 \left(\frac{g^{-1}(t) - y}{h} \right) a(y) dy$$

$$= 4 \int_{-\infty}^{\infty} A(g^{-1}(t) - hu) V(u) K(u) du$$

$$= 2\bar{\mathbb{S}}_X(t) - 2hg'(g^{-1}(t)) \mathbb{S}_X(t) \int_{-\infty}^{\infty} V(u) W(u) du + o(h).$$

And the second term can be calculated with

$$\frac{2}{h} \int_{-\infty}^{\infty} \mathbb{V}_{1,h}(g^{-1}(t), y) K \left(\frac{g^{-1}(t) - y}{h} \right) a(y) dy$$

$$= 2 \int_{-\infty}^{\infty} \left[g'(y) \mathbb{V} \left(\frac{g^{-1}(t) - y}{h} \right) + o(1) \right] K \left(\frac{g^{-1}(t) - y}{h} \right) a(y) dy$$

$$= 2hg'(g^{-1}(t)) \mathbb{S}_X(t) \int_{-\infty}^{\infty} V(u) W(u) du + o(h).$$

That can easily prove the $Var[\widetilde{\mathbb{S}}_{X,1}(t)]$ formula.

Before going into the calculation of the covariance, consider

$$E \left[\mathbb{V}_{1,h}(g^{-1}(t), g^{-1}(X_1)) V \left(\frac{g^{-1}(t) - g^{-1}(X_1)}{h} \right) \right]$$

$$= \int_{-\infty}^{\infty} \left[g'(y) V^2 \left(\frac{g^{-1}(t) - y}{h} \right) \right.$$

$$\left. + \frac{1}{h} \mathbb{V}_{1,h}(g^{-1}(t), y) K \left(\frac{g^{-1}(t) - y}{h} \right) \right] S_Y(y) dy,$$

which once again is needed to calculate them separately. The first term equals

$$\mathbb{S}_X(t) - hg'g^{-1}(t)) S_X(t) \int_{-\infty}^{\infty} V(u) W(u) du + o(h),$$

while the second term is

$$hg'(g^{-1}(t)) S_X(t) \int_{-\infty}^{\infty} V(u) W(u) du + o(h),$$

which will complete the proof.

$\square$

Remark 4.2 Because $\frac{d}{dt} \widetilde{\mathbb{S}}_{X,1}(t) = -\widetilde{S}_X(t)$, it means that the first set of estimators preserves the relationship between the theoretical $\mathbb{S}_X$ and S_X.

It can be seen that $\widetilde{S}_X(t)$ is also basically just a result of a simple substitution of $g^{-1}(t)$ and $g^{-1}(X_i)$ to the formula of $\widehat{S}_X(t)$. This can be done due to the change-of-variable property of the survival function (see Lemma 4.2 and Remark 4.1). Though it is a bit trickier, the change-of-variable property of the cumulative survival function leads to the construction of the second proposed cumulative survival function estimator, which is

$$\tilde{\mathbb{S}}_{X,2}(t) = \frac{1}{n} \sum_{i=1}^{n} \mathbb{V}_{2,h}(g^{-1}(t), g^{-1}(X_i)), \quad t \in \Omega, \tag{4.17}$$

where

$$\mathbb{V}_{2,h}(x, y) = \int_{-\infty}^{y} g'(z) V\left(\frac{x - z}{h}\right) dz. \tag{4.18}$$

In the above formula, multiplying V with g' is necessary to make sure that $\tilde{\mathbb{S}}_{X,2}$ is an estimator of $\mathbb{S}_X$ (see Eq. (4.5)). Its bias and variance are as follows.

Theorem 4.2 *Under conditions C1–C6, the bias and the variance of* $\tilde{\mathbb{S}}_{X,2}(t)$ *are*

$$Bias[\tilde{\mathbb{S}}_{X,2}(t)] = \frac{h^2}{2} b_3(t) \int_{-\infty}^{\infty} y^2 K(y) dy + o(h^2), \tag{4.19}$$

$$Var[\tilde{\mathbb{S}}_{X,2}(t)] = \frac{1}{n} [2\tilde{\mathbb{S}}_X(t) - \mathbb{S}_X^2(t)] + o\left(\frac{h}{n}\right), \tag{4.20}$$

where

$$b_3(t) = [g'(g^{-1}(t))]^2 f_X(t) - g''(g^{-1}(t)) S_X(t). \tag{4.21}$$

Furthermore,

$$Cov[\tilde{\mathbb{S}}_{X,2}(t), \tilde{S}_X(t)] = \frac{1}{n} \mathbb{S}_X(t) F_X(t) + o\left(\frac{h}{n}\right). \tag{4.22}$$

Proof Utilizing similar reasoning as before, then

$$E[\tilde{\mathbb{S}}_{X,2}(t)] = \int_{-\infty}^{\infty} g'(y) V\left(\frac{g^{-1}(t) - y}{h}\right) S_Y(y) dy$$

$$= \mathbb{S}_X(t) + \frac{h^2}{2} b_3(t) \int_{-\infty}^{\infty} u^2 K(u) du + o(h^2),$$

and this proves the bias part. For the variance, consider

$$E[\mathbb{V}_{2,h}^2(g^{-1}(t), g^{-1}(X_1))] = 2 \int_{-\infty}^{\infty} \left[g'(y) V^2 \left(\frac{g^{-1}(t) - y}{h} \right) \right.$$
$$\left. + \frac{1}{h} \mathbb{V}_{2,h}(g^{-1}(t), y) K \left(\frac{g^{-1}(t) - y}{h} \right) \right] a(y) \mathrm{d}y.$$

The first term is already calculated, while the second term is

$$\frac{2}{h} \int_{-\infty}^{\infty} \mathbb{V}_{2,h}(g^{-1}(t), y) K \left(\frac{g^{-1}(t) - y}{h} \right) a(y) \mathrm{d}y$$
$$= 2hg'(g^{-1}(t)) \mathbb{S}_X(t) \int_{-\infty}^{\infty} V(u) W(u) \mathrm{d}u + o(h).$$

Then, the formula of $Var[\widetilde{\mathbb{S}}_{X,2}(t)]$ can be proven. The calculation of the covariance is similar to $Cov[\widetilde{\mathbb{S}}_{X,1}(t), \widetilde{S}_X(t)]$.

Remark 4.3 In Theorems 4.1 and 4.2, a lot of similarities are possessed by both sets of estimators. For example, they have the same covariances, which means the statistical relationship between $\widetilde{S}_X$ and $\widetilde{\mathbb{S}}_{X,2}$ is same to the one of $\widetilde{S}_X$ and $\widetilde{\mathbb{S}}_{X,1}$.

Remark 4.4 It is clear that $\widetilde{S}_X(a_1) = 1$ and it is obvious that $\widetilde{S}_X(a_2) = 0$. Hence, it is clear that their variances are 0 when t approaches the boundaries. This is one of the reasons that the proposed methods outperform the naive kernel estimator.

4.2 Estimators of the Mean Residual Life Function

In this section, the estimation of the mean residual life function will be discussed. As the survival function and the cumulative survival function estimators already have been defined, plugging them into the MRL function formula is clear. Hence, the proposed estimators of the mean excess loss function are

$$\widetilde{m}_{X,1}(t) = \frac{\widetilde{\mathbb{S}}_{X,1}(t)}{\widetilde{S}_X(t)} = \frac{\sum_{i=1}^{n} \int_{g^{-1}(t)}^{\infty} g'(z) V \left(\frac{z - g^{-1}(X_i)}{h} \right) \mathrm{d}z}{\sum_{i=1}^{n} V \left(\frac{g^{-1}(t) - g^{-1}(X_i)}{h} \right)}, \quad t \in \Omega, \qquad (4.23)$$

and

$$\widetilde{m}_{X,2}(t) = \frac{\widetilde{\mathbb{S}}_{X,2}(t)}{\widetilde{S}_X(t)} = \frac{\sum_{i=1}^{n} \int_{-\infty}^{g^{-1}(X_i)} g'(z) V \left(\frac{g^{-1}(t) - z}{h} \right) \mathrm{d}z}{\sum_{i=1}^{n} V \left(\frac{g^{-1}(t) - g^{-1}(X_i)}{h} \right)}, \quad t \in \Omega. \qquad (4.24)$$

At the first glance, $\widetilde{m}_{X,1}$ seems more representative to the theoretical m_X, because the analytic relationship between S_X and $\mathbb{S}_X$ are preserved by $\widetilde{S}_X$ and $\widetilde{\mathbb{S}}_{X,1}$, as stated in Remark 4.2. This is not a major problem for $\widetilde{m}_{X,2}$, as Remark 4.3 explained that the relationship between $\widetilde{S}_X$ and $\widetilde{\mathbb{S}}_{X,2}$ is statistically same to the relationship between $\widetilde{S}_X$ and $\widetilde{\mathbb{S}}_{X,1}$. However, when a statistician wants to keep the analytic relationship between the survival and the cumulative survival functions in their estimates, it is suggested to use $\widetilde{m}_{X,1}$ instead.

Theorem 4.3 *Under conditions C1 until C6, the biases and the variances of* $\widetilde{m}_{X,i}(t)$, *$i = 1, 2$, are*

$$Bias[\widetilde{m}_{X,1}(t)] = \frac{h^2}{2S_X(t)}[b_2(t) + m_X(t)b_1(t)] \int_{-\infty}^{\infty} y^2 K(y) \mathrm{d}y + o(h^2), \quad (4.25)$$

$$Bias[\widetilde{m}_{X,2}(t)] = \frac{h^2}{2S_X(t)}[b_3(t) + m_X(t)b_1(t)] \int_{-\infty}^{\infty} y^2 K(y) \mathrm{d}y + o(h^2), \quad (4.26)$$

$$Var[\widetilde{m}_{X,i}(t)] = \frac{1}{n}\frac{b_4(t)}{S_X^2(t)} - \frac{h}{n}\frac{b_5(t)}{S_X^2(t)} \int_{-\infty}^{\infty} V(y)W(y)\mathrm{d}y + o\left(\frac{h}{n}\right), \quad (4.27)$$

where

$$b_4(t) = 2\bar{\mathbb{S}}_X(t) - S_X(t)m_X^2(t) \quad \text{and} \quad b_5(t) = g'(g^{-1}(t))f_X(t)m_X^2(t). \quad (4.28)$$

Proof For a fixed t, it is clear that $\widetilde{S}_X(t)$ and $\widetilde{\mathbb{S}}_{X,1}(t)$ are consistent estimators for $S_X(t)$ and $\mathbb{S}_X(t)$, respectively, then

$$\widetilde{m}_{X,1}(t) - m_X(t) = \frac{\widetilde{\mathbb{S}}_{X,1}(t) - \widetilde{S}_X(t)m_X(t)}{S_X(t)}\left[1 + \frac{S_X(t) - \widetilde{S}_X(t)}{\widetilde{S}_X(t)}\right]$$

$$= \frac{\widetilde{\mathbb{S}}_{X,1}(t) - \widetilde{S}_X(t)m_X(t)}{S_X(t)}[1 + o_p(1)].$$

Thus, using Theorem 4.1, hence

$$Bias[\widetilde{m}_{X,1}(t)] = \frac{1}{S_X(t)}[E\{\widetilde{\mathbb{S}}_{X,1}(t)\} - m_X(t)E\{\widetilde{S}_X(t)\}]$$

$$= \frac{h^2}{2S_X(t)}[b_2(t) + m_X(t)b_1(t)] \int_{-\infty}^{\infty} y^2 K(y)\mathrm{d}y + o(h^2).$$

The same argument easily proves the formula of $Bias[\widetilde{m}_{X,2}(t)]$.

Using a similar method, for $i = 1, 2$, then

$$Var[\widetilde{m}_{X,i}(t)] = Var\left[\frac{\widetilde{\mathbb{S}}_{X,i}(t) - \widetilde{S}_X(t)m_X(t)}{S_X(t)}\right]$$

$$= \frac{1}{n}\frac{b_4(t)}{S_X^2(t)} - \frac{h}{n}\frac{b_5(t)}{S_X^2(t)} \int_{-\infty}^{\infty} V(y)W(y)\mathrm{d}y + o\left(\frac{h}{n}\right). \qquad \square$$

Remark 4.5 It can be seen from the bias formulas, both the naive kernel MRL estimator and the proposed estimators are comparable in the interior. However, since the bijective transformation successfully eliminates the boundary problem, the biases of the proposed estimators are still in the order of h^2 even in the boundary regions. It is totally different for naive kernel method, where its bias changes the order from h^2 to h or 1. This analytically explains why the proposed method will outperform the naive kernel one in practice.   □

Similar to most of the kernel-type estimators, the proposed estimators attain asymptotic normality and uniformly strong consistency, as stated as follows.

Theorem 4.4 *Under conditions C1–C6, the limiting distribution*

$$\frac{\tilde{m}_{X,i}(t) - m_X(t)}{\sqrt{Var[\tilde{m}_{X,i}(t)]}} \to_D N(0, 1)$$

holds for $i = 1, 2$.

Proof Because the proof of the case $i = 1$ is similar, only the case of $i = 2$ will be provided in detail. First, for some $\delta > 0$, using Hölder and c_r inequalities results in

$$E\left[\left|V\left(\frac{g^{-1}(t) - g^{-1}(X_1)}{h}\right) - E\left\{V\left(\frac{g^{-1}(t) - g^{-1}(X_1)}{h}\right)\right\}\right|^{2+\delta}\right]$$

$$\leq 2^{2+\delta} E\left[\left|V\left(\frac{g^{-1}(t) - g^{-1}(X_1)}{h}\right)\right|^{2+\delta}\right].$$

But, since $0 \leq V(\cdot) \leq 1$, then

$$E\left[\left|V\left(\frac{g^{-1}(t) - g^{-1}(X_1)}{h}\right) - E\left\{V\left(\frac{g^{-1}(t) - g^{-1}(X_1)}{h}\right)\right\}\right|^{2+\delta}\right] \leq 2^{2+\delta} < \infty,$$

and because $Var[V\{(g^{-1}(t) - g^{-1}(X_1))/h\}] = O(1)$, then

$$\frac{E\left[\left|V\left(\frac{g^{-1}(t)-g^{-1}(X_1)}{h}\right) - E\left\{V\left(\frac{g^{-1}(t)-g^{-1}(X_1)}{h}\right)\right\}\right|^{2+\delta}\right]}{n^{\delta/2}\left[Var\left\{V\left(\frac{g^{-1}(t)-g^{-1}(X_1)}{h}\right)\right\}\right]^{1+\delta/2}} \to 0$$

when $n \to \infty$. Hence, with the fact $\tilde{S}_X(t) \to_p S_X(t)$, it can be concluded that

$$\frac{\tilde{S}_X(t) - S_X(t)}{\sqrt{Var[\tilde{S}_X(t)]}} \to_D N(0, 1)$$

holds [38]. Next, with a similar reasoning as before,

$$E[|\mathbb{V}_{2,h}(g^{-1}(t), g^{-1}(X_1)) - E\{\mathbb{V}_{2,h}(g^{-1}(t), g^{-1}(X_1))\}|^{2+\delta}]$$
$$\leq 2^{2+\delta} E[|\mathbb{V}_{2,h}(g^{-1}(t), g^{-1}(X_1))|^{2+\delta}],$$

which, by the same inequalities, results in

$$E[|\mathbb{V}_{2,h}(g^{-1}(t), g^{-1}(X_1)) - E\{\mathbb{V}_{2,h}(g^{-1}(t), g^{-1}(X_1))\}|^{2+\delta}]$$
$$\leq 2^{2+\delta} E\left[\left|\int_{-\infty}^{g^{-1}(X_1)} g'(z) V\left(\frac{g^{-1}(t) - z}{h}\right) dz\right|^{2+\delta}\right]$$
$$\leq 2^{2+\delta} E\left[\left|\int_{-\infty}^{g^{-1}(X_1)} g'(z) dz\right|^{2+\delta}\right]$$
$$\leq 2^{2+\delta} E(X_1^{2+\delta})$$
$$< \infty.$$

Therefore, the same argument implies

$$\frac{\widetilde{\mathbb{S}}_{X,2}(t) - \mathbb{S}_X(t)}{\sqrt{Var[\widetilde{\mathbb{S}}_{X,2}(t)]}} \to_D N(0, 1).$$

At last, by Slutsky's Theorem for rational function, the theorem is proven. Note that $\delta \leq 1$. ∎

Theorem 4.5 *Under conditions C1–C6, the consistency*

$$\sup_{t \in \Omega} |\widetilde{m}_{X,i}(t) - m_X(t)| \to_{a.s.} 0$$

holds for $i = 1, 2$.

Proof It is guaranteed that $\sup_{t \in \mathbb{R}} |\widehat{S}_Y(t) - S_Y(t)| \to_{a.s.} 0$ [2], which implies

$$\sup_{t \in \Omega} |\widehat{S}_Y(g^{-1}(t)) - S_Y(g^{-1}(t))| \to_{a.s.} 0.$$

However, because $S_Y(g^{-1}(t)) = S_X(t)$ and $\widehat{S}_Y(g^{-1}(t)) = \widetilde{S}_X(t)$, then $\sup_{t \in \Omega} |\widetilde{S}_X(t) - S_X(t)| \to_{a.s.} 0$ holds.

Next, since $\mathbb{S}_X(t) \geq 0$ is bounded above with

$$\sup_{t \in \Omega} \mathbb{S}_X(t) = \lim_{t \to a_1^+} \mathbb{S}_X(t) = E(X) - a_1,$$

then $a(g^{-1}(t)) = \mathbb{S}_X(t)$ is bounded on Ω. Furthermore,

$$\widehat{a}(g^{-1}(t)) = \frac{1}{n} \sum_{i=1}^{n} \int_{t}^{a_2} V\left(\frac{g^{-1}(z) - g^{-1}(X_i)}{h}\right) dz = \widetilde{\mathbb{S}}_{X,1}(t)$$

is also bounded above almost surely with

$$\sup_{t \in \Omega} \widetilde{\mathbb{S}}_{X,1}(t) = \lim_{t \to a_1^+} \widetilde{\mathbb{S}}_{X,1}(t) = \bar{X} - a_1 + O_p(h^2).$$

Thus, Lemma 4.3 implies

$$\sup_{t \in \Omega} |\widetilde{\mathbb{S}}_{X,1}(t) - \mathbb{S}_X(t)| = \sup_{t \in \Omega} |\widehat{a}(g^{-1}(t)) - a(g^{-1}(t))| \to_{a.s.} 0.$$

As a conclusion, $\sup_{t \in \Omega} |\widetilde{m}_{X,1}(t) - m_X(t)| \to_{a.s.} 0$ holds. The proof for $i = 2$ is similar.   □

Remark 4.6 A construction of point-wise confidence intervals for the MRL function using the proposed estimators by utilizing normal approximation is possible, as implied by Theorem 4.4. Moreover, since the proposed methods are also uniformly consistent (Theorem 4.5), it is also possible to define its confidence bands.

The last property that will be discussed is the behavior of the proposed estimators when t is in the boundary regions. As stated in the introduction, estimators that preserve the behavior of the theoretical MRL function are of interest, specifically the mean value property. If this can be proven, then not only will the proposed methods be free of boundary problems but also superior in the sense of preserving the key property of the MRL function.

Theorem 4.6 *Let $\widetilde{m}_{X,1}$ and $\widetilde{m}_{X,2}$ be the proposed boundary-free kernel mean residual life function estimators. Then*

$$\widetilde{m}_{X,1}(a_1) + a_1 = \bar{X} + O_p(h^2) \tag{4.29}$$

and

$$\widetilde{m}_{X,2}(a_1) + a_1 = \bar{X}. \tag{4.30}$$

Proof To calculate $\lim_{t \to a_1^+} \widetilde{m}_{X,i}(t)$, see the limit behavior of each estimator of the survival function and the cumulative survival function. First, the fact $\lim_{x \to -\infty} V(x) = 1$ results in

$$\lim_{t \to a_1^+} \widetilde{S}_X(t) = \frac{1}{n} \sum_{i=1}^{n} \lim_{t \to a_1^+} V\left(\frac{g^{-1}(t) - g^{-1}(X_i)}{h}\right) = 1.$$

For $\lim_{t \to a_1^+} \widetilde{\mathbb{S}}_{X,1}(t)$, the use of the integration by substitution and by parts means

$$\lim_{t \to a_1^+} \widetilde{\mathbb{S}}_{X,1}(t) = -a_1 + \frac{1}{n} \sum_{i=1}^{n} \int_{-\infty}^{\infty} g(g^{-1}(X_i) + hu) K(u) du$$

$$= \frac{1}{n} \sum_{i=1}^{n} \int_{-\infty}^{\infty} [g(g^{-1}(X_i)) + hg'(g^{-1}(X_i))u + O_p(h^2)] K(u) du - a_1$$

$$= \bar{X} - a_1 + O_p(h^2).$$

At last,

$$\lim_{t \to a_1^+} \widetilde{\mathbb{S}}_{X,2}(t) = \frac{1}{n} \sum_{i=1}^{n} \int_{-\infty}^{g^{-1}(X_i)} g'(z) \lim_{t \to a_1^+} V\left(\frac{g^{-1}(t) - z}{h}\right) dz = \bar{X} - a_1.$$

Hence, the theorem is proven. □

Remark 4.7 Note that, although for convenience it is written as $\widetilde{m}_{X,i}(a_1)$, actually it means as $\lim_{t \to a_1^+} \widetilde{m}_{X,i}(t)$, since $g^{-1}(a_1)$ might be undefined.

Remark 4.8 From Eq. (4.30), $\widetilde{m}_{X,2}(a_1)$ is an unbiased estimator, because

$$E[\widetilde{m}_{X,2}(a_1)] = E(X) - a_1 = m_X(a_1).$$

In other words, its bias is exactly 0. On the other hand, though $\widetilde{m}_{X,1}(a_1)$ is not exactly the same as $\bar{X} - a_1$, at least it is reasonable to say that they are close enough, and the rate of h^2 error is relatively small. However, from this, it may be concluded that $\widetilde{m}_{X,2}$ is superior to $\widetilde{m}_{X,1}$ in the aspect of preserving behavior of the MRL function near the boundary.

From the preceding expositions, the root of difference between the proposed estimators is in their constructions. The first estimator of the cumulative survival function is constructed by $\widetilde{\mathbb{S}}_{X,1}(t) = \int_t^{a_2} \widetilde{S}_X(y) dy$, which mimics the theoretical $\mathbb{S}_X(t) = \int_t^{a_2} S_X(x) dx$. This property may appeal to some statisticians who want to preserve such analytical relationship in their estimation of m_X. Also, from the numerical point of view, calculating $\widetilde{m}_{X,1}$ is slightly easier, as applying numerical integration technique on $\widetilde{S}_X$ for computing $\widetilde{\mathbb{S}}_{X,1}$ is easy.

On the other hand, the second estimator of the cumulative survival function is constructed by $\widetilde{\mathbb{S}}_{X,2}(t) = \int_t^{a_2} g'(x) \widetilde{S}_X(x) dx$, which mimics the theoretical change-of-variable property (see Eq. (4.5)). This might make the first estimator more intuitive, as the second estimator does not enjoy the integration relationship naturally owned by the mean residual life function. However, as stated in Remark 4.8, the estimator $\widetilde{m}_{X,2}$ analytically performs better in the boundary region due to its preservation of mean value property.

4.3   Numerical Studies

In this section, the results of the numerical studies will be shown. The studies are divided into two parts, the simulations, and the real data analysis.

The average integrated squared error (AISE) is calculated with several sample sizes, and repeated them 1000 times for each case. Five estimators will be compared: empirical m_n, modified empirical likelihood (MEL) [33], naive kernel $\widehat{m}_X$, and the two proposed estimators. For the proposed methods, two mappings g^{-1} were chosen for each case. When $\Omega = \mathbb{R}^+$, a [a]composite of two functions $\Phi^{-1} \circ \gamma$, where $\gamma(x) = 1 - e^{-x}$, and the [b]logarithm function log, were used. In the case of $\Omega = [0, 1]$, [a]logit function Π^{-1}, where Π is the logistic distribution function, and [b]probit function Φ^{-1}, were applied.

The distributions the generated data are exponential $\exp(1)$, gamma $\Gamma(2, 3)$, Weibull $Wei(3, 2)$, standard absolute-normal $abs.N(0, 1)$, standard uniform $U(0, 1)$, also beta distributions ($\beta(3, 3)$, $\beta(2, 4)$, and $\beta(4, 2)$). The gamma and absolute-normal distributions were chosen because they are lighter tailed, while exponential and Weibull distributions have moderate and heavier tails, respectively. Furthermore, the beta distribution with various parameters was simulated as they represent right-skewed, left-skewed, and symmetric distributions, while the uniform distribution was chosen as it has flat probability. The kernel function used here is the Epanechnikov one and the bandwidths were chosen by cross-validation.

In Table 4.1, the proposed estimators gave the best results for all cases. This is particularly true for the second proposed estimator when the support of the data is nonnegative real line, which logarithm function worked better. However, since the differences are not so large, $\Phi^{-1} \circ \gamma$ still can be a good choice. On the other hand, when the support is the unit interval, the first proposed method seemed to work slightly better than the second one. Though, in general, the first proposed estimator's performances are not as good as the second one, it is still fairly comparable.

As further illustrations, some graphs to compare the point-wise performances of the proposed estimators to the other estimators are also provided. Figures 4.1a and 4.2a illustrate the point-wise estimates of each estimator from several distributions (without repetition). Furthermore, the other graphs are depicting the point-wise simulated bias, simulated variance, and the average squared error (ASE) of each estimator after 1000 repetitions. From these figures, there are several things needed to be emphasized.

First, one might think that the empirical MRL function works the best because its simulated bias lines are almost always the nearest to 0 (Figs. 4.1b and 4.2b). This is an obvious result because the empirical one is an unbiased estimator. However, the unbiasedness should be paid by the lack of smoothness, as all the point-wise estimates are so jumpy. Moreover, lack of smoothness will cause instability of estimation, which can be seen by how big the variance values of the empirical estimator are in Fig. 4.1c and 4.2c.

Second, all figures show how bad the performances of the naive kernel mean residual life function estimator. It means the boundary bias does affect the performances

Table 4.1 Average integrated squared error ($\times 10^5$) comparison

n	Distribution	Empirical	MEL	Naive	Prop. 1[a]	Prop. 1[b]	Prop. 2[a]	Prop. 2[b]
	$Exp(1)$	132765	136064	229308	68007	76069	63148	**52140**
	$\Gamma(2,3)$	704072	612704	1685292	411047	546663	285871	**228580**
	$Wei(3,2)$	56464	56799	70314	30828	55582	**4987**	7477
50	$abs.N(0,1)$	18951	18744	32179	11507	18141	4819	**4204**
	$U(0,1)$	865	806	1255	645	**628**	732	777
	$\beta(3,3)$	582	593	1310	**411**	520	443	577
	$\beta(2,4)$	923	832	1069	562	717	**527**	586
	$\beta(4,2)$	606	617	1486	**349**	387	440	586
	$Exp(1)$	71433	71521	115213	34122	43125	32024	**31021**
	$\Gamma(2,3)$	411131	401234	843151	201114	322342	143440	**124351**
	$Wei(3,2)$	33232	30131	41102	21314	23441	**2544**	4234
100	$abs.N(0,1)$	9530	9135	21045	6313	9020	2405	**2111**
	$U(0,1)$	434	441	634	323	**314**	421	345
	$\beta(3,3)$	351	345	701	**201**	311	232	343
	$\beta(2,4)$	512	523	535	331	405	**314**	347
	$\beta(4,2)$	312	321	743	**225**	244	229	309
	$Exp(1)$	30212	30213	52101	12012	21012	11122	**10120**
	$\Gamma(2,3)$	201210	201112	421020	112342	111121	71231	**62120**
	$Wei(3,2)$	19111	19197	20120	10102	11220	**1229**	2112
200	$abs.N(0,1)$	4211	4217	10122	3101	4101	1211	**1017**
	$U(0,1)$	213	213	312	111	**102**	210	122
	$\beta(3,3)$	120	127	316	**110**	123	112	121
	$\beta(2,4)$	200	205	212	110	211	**101**	123
	$\beta(4,2)$	101	111	321	**95**	112	114	115

of the naive kernel estimator. In particular, in Fig. 4.2, glaring gaps can be seen between the naive kernel estimator and the others. It is because when $\Omega = [0,1]$, there are two boundary points, which means they double up the boundary problems.

Third, when t is near to 0, the second proposed estimator and the empirical one work the best. All the MRL estimates and simulated bias figures show that the values of $\widetilde{m}_{X,2}(0)$ and $m_n(0)$ are the closest to $m_X(0)$. Also, it is clear that simulated $Bias[\widetilde{m}_{X,2}(0)]$ and $Bias[m_n(0)]$ are both 0 (Theorem 4.6). Though the performances of the first proposed method are not as good when t is near to 0, but the gaps are really small, meaning the $O_p(h^2)$ error is good enough.

Lastly, it is clear from the simulated variance and ASE, that generally the proposed estimators work the best. It is also worth to note that, when $\Omega = \mathbb{R}^+$, all the point-wise estimates presented here will fade to 0 when t is large enough, but the proposed estimators are more stable and fading to 0 much slower than the other two estimators. Conversely, when $\Omega = [0,1]$, the naive kernel estimates fail to reach 0 even when $t = 1$, where the empirical and the proposed methods follow the theoretical values truthfully.

For real data analysis, the UIS Drug Treatment Study Data [39] were analyzed to show the performances of the proposed methods for real data. The dataset records the

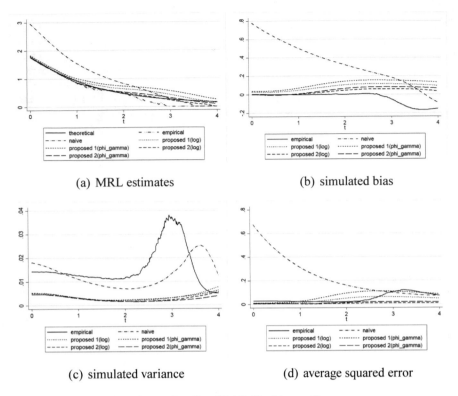

(a) MRL estimates   (b) simulated bias

(c) simulated variance   (d) average squared error

Fig. 4.1 Point-wise comparisons when $X \sim Wei(3, 2)$ with $n = 50$

result of an experiment about how long someone who got drug treatment to relapse (reuse) the drug again. The variable used in the calculation is the "time" variable, which represents the number of days after the admission to drug treatment until drug relapse.

Figure 4.3 shows that, once again, the naive kernel estimator is just a smoothed version of the empirical MRL function. Furthermore, soon after m_n touches 0, $\widehat{m}_X$ also reaches 0. Conversely, though the proposed estimators are decreasing as well, but they are much slower than the other two, and they are much smoother than the others.

Theoretically, any transformation g^{-1} can be chosen as long as $g : \mathbb{R} \to \Omega$ is bijective, but here a suggestion as a guidance in real-life settings was written. First, do an explanatory data analysis to determine the basic quantities of the data, especially its skewness. If the support of the data is a bounded interval $[a, b]$, transforming the data first to $[0, 1]$ with formula $d(x) = (x - a)/(b - a)$ is suggested. Now, if the data is highly skewed, choose logit function to transform the data to be more symmetric (hence the normality stated in Theorem 4.4 would converge faster), then the transformation to apply to the proposed method would be $g^{-1} = \Pi^{-1} \circ d$. Otherwise, if the data is somewhat symmetric, choose $g^{-1} = \Phi^{-1} \circ d$. When the support

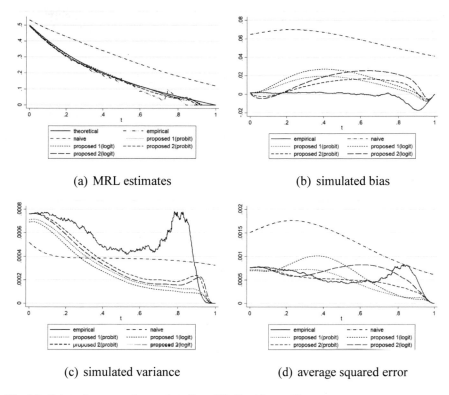

(a) MRL estimates   (b) simulated bias

(c) simulated variance   (d) average squared error

Fig. 4.2 Point-wise comparisons when $X \sim \beta(3, 3)$ with $n = 50$

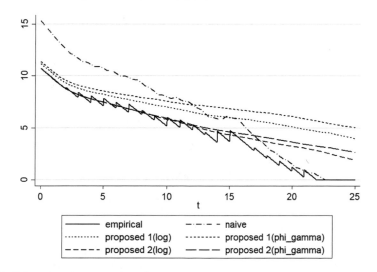

Fig. 4.3 Point-wise wise comparison for UIS data

of the data is $[a, \infty)$, use $g^{-1}(x) = \log(x - a)$ transformation if it is highly skewed, and choose $g^{-1}(x) = \Phi^{-1}(1 - e^{a-x})$ if the data is symmetric enough.

The above suggestion is written based on the common concept in Econometrics that, when the data is far from "normal", logarithmic transformation would help to normalize it. This suggestion is also illustrated in Table 4.1, as the logarithmic and logit transformations worked better on exponential, gamma, absolute-normal, and beta distributions. However, please note that this suggestion is *not universally true*, as sometimes probit transformation works better even for skewed data. Note that there is no restriction to use probit, logit, and logarithmic transformations only, as other transformations, such as the inverse of Gumbel or Student's t Distribution functions, are also applicable. Regardless, the choice of transformation and the choice of which proposed estimator to use do not change the result drastically.

References

1. Embrechts P, Klüppelberg C, Mikosch T (1997) Modelling extremal events. Springer, Berlin
2. Belzunce F, Ruiz JM, Pellerey F, Shaked M (1996) The dilation order, the dispersion order, and orderings of residual lifes. Stat Probab Lett 33:263–275
3. Zoroa P, Ruiz JM, Marlin J (1990) A characterization based on conditional expectations. Commun Stat Theory Method 19:3127–3135
4. Ruiz JM, Navarro J (1994) Characterization of distributions by relationships between the failure rate and the mean residual life. IEEE Trans Reliab 43:640–644
5. Navarro J, Hernandez PJ (2008) Mean residual life functions of finite mixtures, order statistics, and coherent systems. Metr 67:277–298
6. Dileepkumar M, Sankaran PG (2022) On proportional odds relevation transform and its applications. Commun Stat Theory Method. https://doi.org/10.1080/03610926.2022.2129994
7. Nanda AK, Bhattacharjee S, Balakrishnan N (2010) Mean residual life function, associated orderings, and properties. IEEE Trans Reliab 59(1):55–65
8. Huynh KT, Castro IT, Barros A, Bérenguer C (2014) On the use of mean residual life as a condition index for condition-based maintenance decision-making. IEEE Trans Syst Man Cyber Syst 44(7):877–893
9. Sadegh MK (2008) Mean past and mean residual life functions of a parallel system with nonidentical components. Commun Stat Theory Method 37(7):1134–1145
10. Raqab MZ, Asadi M (2008) On the mean residual life of records. J Stat Plan Inference 138:3660–3666
11. Eryilmaz S (2012) On the mean residual life of a k-out-of-n: G system with a single cold standby component. Eur J Oper Res 222:273–277
12. Poursaeed MH (2010) A note on the mean past and the mean residual life of a $(n - k + 1)$-out-of-n system under multi monitoring. Stat Pap 51:409–419
13. Eryilmaz S (2017) Computing optimal replacement time and mean residual life in reliability shock models. Comput Ind Eng 103:40–45
14. Bayramoglu I, Ozkut M (2016) Mean residual life and inactivity time of a coherent system subjected to Marshall-Olkin type shocks. J Comput Appl Math 298:190–200
15. Eryilmaz S, Coolen FPA, Coolen-Maturi T (2018) Mean residual life of coherent systems consisting of multiple types of dependent components. Nav Res Logist 65(1):86–97
16. Zhao S, Makis V, Chen S, Yong L (2018) Evaluation of reliability function and mean residual life for degrading systems subject to condition monitoring and random failure. IEEE Trans Reliab 67(1):13–25

17. Ghasemi A, Hodkiewicz MR (2012) Estimating mean residual life for a case study of rail wagon bearings. IEEE Trans Reliab 61(3):719–730
18. Kim B, Kim J, Lee S (2022) Joint pricing and inventory control for a production-inventory queueing system. Ann Oper Res. https://doi.org/10.1007/s10479-022-04948-1
19. Rahimdel MJ (2022) Residual lifetime estimation for the mining truck tires. Proc Inst Mech Eng D J Automob Eng. https://doi.org/10.1177/09544070221121855
20. Csörgő M, Zitikis R (1996) Mean residual life processes. Ann Stat 24:1717–1739
21. Sun L, Zhang Z (2009) A class of transformed mean residual life models with censored survival data. J Am Stat Assoc 104(486):803–815
22. Shen Y, Tang LC, Xie M (2009) A model for upside-down bathtub-shaped mean residual life and its properties. IEEE Trans Reliab 58(3):425–431
23. Zhao P, Balakrishnan N (2009) Mean residual life order of convolutions of heterogeneous exponential random variables. J Multivar Anal 100:1792–1801
24. Chan KCG, Chen YQ, Di CZ (2012) Proportional mean residual life model for right-censored length-biased data. Biom 99(4):995–1000
25. Sun L, Song X, Zhang Z (2012) Mean residual life models with time-dependent coefficients under right censoring. Biom 99(1):185–197
26. Ocloo SK, Brew L, Nasiru S, Odoi B (2022) Harmonic mixture Fréchet distribution: properties and applications to lifetime data. Int J Math Math Sci. https://doi.org/10.1155/2022/6460362
27. Shanker R, Upadhyay R, Shukla KK (2022) A quasi Suja distribution. Reliab Theory Appl 17(3(69)):162–178
28. Sah BK, Sahani SK (2022) Polynomial-exponential distribution. Math Stat Eng Appl 71(4):2474–2486
29. Aljuhani W, Klakattawi HS, Baharith LA (2022) Alpha power exponentiated new Weibull-Pareto distribution: its properties and applications. Pak J Stat Oper Res https://doi.org/10.18187/pjsor.v18i3.3937
30. Rasheed H, Dar IS, Saqib M, Abbas N, Suhail M (2022) The odd modified Burr-III exponential distribution: properties, estimation, and application. J Nat Sci Found Sri Lanka 50(2):425–439
31. Ruiz JM, Guillamón A (1996) Nonparametric recursive estimator for mean residual life and vitality function under dependence conditions. Commun. Stat Theory Method 25:1997–2011
32. Chaubey YP, Sen PK (1999) On smooth estimation of mean residual life. J Stat Plan Inference 75:223–236
33. Ratnasingam S, Ning W (2022) Confidence intervals of mean residual life function in length-biased sampling based on modified empirical likelihood. J Biopharma Stat. https://doi.org/10.1080/10543406.2022.2089157
34. Guillamón A, Navarro J, Ruiz JM (1998) Nonparametric estimator for mean residual life and vitality function. Stat Pap 39:263–276
35. Fauzi RR, Maesono Y (2023) Boundary-free estimators of the mean residual life function for data on general interval. Commun Stat Theory Method. https://doi.org/10.1080/03610926.2023.2168484
36. Geenens G (2014) Probit transformation for kernel density estimation on the unit interval. J Am Stat Assoc 109:346–358
37. Fauzi RR, Maesono Y (2021) Boundary-free kernel-smoothed goodness-of-fit tests for data on general interval. Commun Stat Simul Comput. https://doi.org/10.1080/03610918.2021.1894336
38. Loeve M (1963) Probability theory. Van Nostrand-Reinhold, New Jersey
39. Hosmer DW, Lemeshow S (1998) Applied survival analysis: Regression modeling of time to event data. John Wiley and Sons, New York

Chapter 5
Kernel-Based Nonparametric Tests

Abstract Besides estimation, another important aspect of statistical inference is statistical testing. In this chapter, kernel-smoothed version of three well-known statistical tests are introduced. The three tests consist of Kolmogorov-Smirnov, Cramér-von Mises, and Wilcoxon signed test. Though the distributions of their test statistics converge to the same distributions as their unsmoothed counterpart, their improvement in minimizing errors can be proven. Some simulation results illustrating the estimator and the tests' performances will be presented in the last part of this article.

As stated in the name, parametric statistical methods depend on a distributional assumption of the data under consideration. For example, tests of normality for residuals are needed to justify the fulfillment of normality assumption in the linear regression. In most cases, assumption about the distribution of the data dictates the methods that can be used to estimate the parameters and also determines the methods that statisticians could apply. Some examples of goodness-of-fit tests are the Kolmogorov-Smirnov (KS) test, Cramér-von Mises (CvM) test, Anderson-Darling test, and Durbin-Watson test. In this chapter, the KS and CvM tests are the focus.

Let $X_1, X_2, \ldots, X_n$ be independently and identically distributed random variables supported on $\Omega \subseteq \mathbb{R}$ with an absolutely continuous distribution function F_X and a density f_X. In this setting, the Kolmogorov-Smirnov statistic utilizes the empirical distribution function F_n to test the null hypothesis

$$H_0 : F_X = F$$

against the alternative hypothesis

$$H_1 : F_X \neq F,$$

where F is the assumed distribution function. The test statistic is defined as

$$K S_n = \sup_{x \in \mathbb{R}} | F_n(x) - F(x) |. \tag{5.1}$$

If under a significance level α the value of KS_n is larger than a certain value from the Kolmogorov distribution table, H_0 is rejected. Likewise, under the same circumstance, the statistic of the Cramér-von Mises test is defined as

$$CvM_n = n \int_{-\infty}^{\infty} [F_n(x) - F(x)]^2 dF(x), \tag{5.2}$$

and the null hypothesis is rejected when the value of CvM_n is larger than a certain value from Cramér-von Mises table.

Several discussions regarding those goodness-of-fit tests have been around for decades. The recent articles include the distribution of KS and CvM tests for exponential populations [1], revision of two-sample KS test [2], CvM distance for neighborhood-of-model validation [3], rank-based CvM test [4], and model selection using CvM distance in a fixed design regression [5].

5.1   Naive Kernel Goodness-of-Fit Tests

The standard tests work really well in many cases, but it does not mean it is without any problem. The sensitivity near the center of distribution caused by lack of smoothness of F_n may increase the probability of type-1 error more than the intended α, especially when the sample size n is small. Also, since the distribution function F_X is absolutely continuous, it seems to be more appropriate to use a smooth and continuous estimator rather than the empirical distribution function F_n for testing the fitness.

Obviously, using the kernel distribution function estimator introduced in Chap. 2 to replace the empirical distribution function is the natural way to smooth the goodness-of-fit statistic out. By that, eliminating the over-sensitivity that standard tests bear is expected. Hence, the formulas become

$$\widehat{KS} = \sup_{x \in \mathbb{R}} |\widehat{F}_X(x) - F(x)| \tag{5.3}$$

and

$$\widehat{CvM} = n \int_{-\infty}^{\infty} [\widehat{F}(x) - F(x)]^2 dF(x), \tag{5.4}$$

where $\widehat{F}_X$ is the (naive) kernel distribution function.

Theorem 5.1 *Let F_X and F be distribution functions on $\mathbb{R}$. Then, under the null hypothesis $F_X = F$,*

$$\left| KS_n - \widehat{KS} \right| \to_p 0$$

and

$$\left| CvM_n - \widehat{CvM} \right| \to_p 0.$$

Proof This theorem is proven by Eq. (16) in [6]. □

By Theorem 5.1, it can be implied that the asymptotic distribution of the kernel Kolmogorov-Smirnov and Cramér-von Mises statistics are the very same Kolmogorov and Cramér distributions, respectively. Hence, the same critical values are used for the test.

5.2 Boundary-Free Kernel-Type Goodness-of-Fit Tests

Though kernel-type goodness-of-fit tests are versatile even when the sample size is relatively small, if the support Ω of the data is strictly smaller than the entire real line, the naive kernel distribution function estimator also suffers the so-called boundary bias problem. As stated in Sect. 1.1, this problem happens because the estimator still puts some weights outside the support Ω. Even though, in some cases (e.g., $f_X(0) = 0$ when 0 is the boundary point), the boundary effects of $\widehat{F}_X(x)$ are not as severe as in the kernel density estimator, the problem still occurs. It is because the value of $\widehat{F}_X(x)$ is still larger than 0 (or less than 1) at the boundary points. This phenomenon cause large value of $|\widehat{F}_X(x) - F(x)|$ in the boundary regions, and then $\widehat{KS}$ and $\widehat{CvM}$ tend to be larger than they are supposed to be, leading to the rejection of H_0 even though H_0 is right.

To solve the boundary bias problem, the same bijective transformation idea as in Sect. 4.1 will be used, then a boundary-free kernel distribution function estimator will be defined. Similarly, some conditions are imposed to make sure this idea is mathematically applicable. The conditions are:

D1. the kernel function $K(v)$ is nonnegative, continuous, and symmetric at $v = 0$,
D2. the integral $\int_{-\infty}^{\infty} v^2 K(v) dv$ is finite and $\int_{-\infty}^{\infty} K(v) dv = 1$,
D3. the bandwidth $h > 0$ satisfies $h \to 0$ and $nh \to \infty$ when $n \to \infty$,
D4. the increasing function g transforms $\mathbb{R}$ onto Ω,
D5. the density f_X and the function g are twice differentiable.

Conditions D1–D3 are standard conditions for kernel method. Albeit it is sufficient for g to be a bijective function, but the increasing property in D4 makes the proofs of the theorems simpler. The last condition is needed to derive the formula of biases and variances.

5.2.1 Boundary-Free KDFE

Under previous conditions, the boundary-free kernel distribution function estimator is defined as

$$\widetilde{F}_X(x) = \frac{1}{n} \sum_{i=1}^{n} W\left(\frac{g^{-1}(x) - g^{-1}(X_i)}{h}\right), \quad x \in \Omega, \tag{5.5}$$

where $h > 0$ is the bandwidth and g is an appropriate bijective function. It can be seen that $\widetilde{F}_X(x)$ is basically just a result of simple substitution of $g^{-1}(x)$ and $g^{-1}(X_i)$ to the formula of $\widehat{F}_X(x)$. Though it looks simple, the argument behind this idea is due to the change-of-variable property of distribution function, which cannot always be done to other probability-related functions. Its bias and variance are given in the following theorem.

Theorem 5.2 *Under conditions D1–D5, the bias and the variance of $\widetilde{F}_X(x)$ are*

$$Bias[\widetilde{F}_X(x)] = \frac{h^2}{2} c_1(x) \int_{-\infty}^{\infty} v^2 K(v)dv + o(h^2), \tag{5.6}$$

$$Var[\widetilde{F}_X(x)] = \frac{1}{n} F_X(x)[1 - F_X(x)] - \frac{2h}{n} g'(g^{-1}(x)) f_X(x) r_1 + o\left(\frac{h}{n}\right), \tag{5.7}$$

where

$$c_1(x) = g''(g^{-1}(x)) f_X(x) + [g'(g^{-1}(x))]^2 f_X'(x). \tag{5.8}$$

Proof Utilizing the usual reasoning of *i.i.d.* random variables and the transformation property of expectation, with $Y = g^{-1}(X_1)$, then

$$E[\widetilde{F}_X(x)] = \frac{1}{h} \int_{-\infty}^{\infty} F_Y(y) K\left(\frac{g^{-1}(x) - y}{h}\right) dy$$

$$= F_X(x) + \frac{h^2}{2} c_1(x) \int_{-\infty}^{\infty} v^2 K(v)dv + o(h^2),$$

and the $Bias[\widetilde{F}_X(x)]$ is obtained. For the variance of $\widetilde{F}_X(x)$, first calculate

$$E\left[W^2\left(\frac{g^{-1}(x) - g^{-1}(X_1)}{h}\right)\right] = \frac{2}{h} \int_{-\infty}^{\infty} F_Y(y) W\left(\frac{g^{-1}(x) - y}{h}\right) K\left(\frac{g^{-1}(x) - y}{h}\right) dy$$

$$= F_X(x) - 2h g'(g^{-1}(x)) f_X(x) r_1 + o(h),$$

and the variance formula is done.   □

Remark 5.1 It is easy to prove that r_1 is a positive number. Then, since g is an increasing function, the variance of the proposed estimator will be smaller than $Var[\widehat{F}_X(x)]$ when $g'(g^{-1}(x)) \geq 1$. On the other hand, though it is difficult to conclude in general case, if the mapping g is carefully chosen, the bias of the proposed method is much faster to converge to 0 than $Bias[\widehat{F}_X(x)]$. For example, when $\Omega = \mathbb{R}^+$ and $g(x) = e^x$ is taken, in the boundary region when $x \to 0$ the bias will converge to 0 faster and $Var[\widetilde{F}_X(x)] < Var[\widehat{F}_X(x)]$.

Similar to most of the kernel-type estimators, the proposed estimator attains asymptotic normality, as stated in the following theorem.

Theorem 5.3 *Under conditions D1–D5, the limiting distribution*

$$\frac{\widetilde{F}_X(x) - F_X(x)}{\sqrt{Var[\widetilde{F}_X(x)]}} \to_D N(0, 1)$$

holds.

Proof For some $\delta > 0$, using Hölder and Cramér c_r inequalities, then

$$E\left[\left|W\left(\frac{g^{-1}(x) - g^{-1}(X_1)}{h}\right) - E\left\{W\left(\frac{g^{-1}(x) - g^{-1}(X_1)}{h}\right)\right\}\right|^{2+\delta}\right]$$

$$\leq 2^{2+\delta} E\left[\left|W\left(\frac{g^{-1}(x) - g^{-1}(X_1)}{h}\right)\right|^{2+\delta}\right].$$

But, since $0 \leq W(v) \leq 1$ for any $v \in \mathbb{R}$, then

$$E\left[\left|W\left(\frac{g^{-1}(x) - g^{-1}(X_1)}{h}\right) - E\left\{W\left(\frac{g^{-1}(x) - g^{-1}(X_1)}{h}\right)\right\}\right|^{2+\delta}\right] \leq 2^{2+\delta} < \infty.$$

Also, because $Var\left[W\left(\frac{g^{-1}(x) - g^{-1}(X_1)}{h}\right)\right] = O(1)$, hence

$$\frac{E\left[\left|W\left(\frac{g^{-1}(x) - g^{-1}(X_1)}{h}\right) - E\left\{W\left(\frac{g^{-1}(x) - g^{-1}(X_1)}{h}\right)\right\}\right|^{2+\delta}\right]}{n^{\delta/2}\left[Var\left\{W\left(\frac{g^{-1}(x) - g^{-1}(X_1)}{h}\right)\right\}\right]^{1+\delta/2}} \to 0$$

when $n \to \infty$. Hence, with the fact $\widetilde{F}_X(x) \to_p F_X(x)$ [12], its asymptotic normality is concluded.   □

Furthermore, strong consistency of the proposed method can be established as well.

Theorem 5.4 *Under conditions D1–D5, the consistency*

$$\sup_{x \in \Omega} |\widetilde{F}_X(x) - F_X(x)| \to_{a.s.} 0$$

holds.

Proof Let F_Y and $\widehat{F}_Y$ be the distribution function and the naive kernel distribution function estimator, respectively, of $Y_1, Y_2, \ldots, Y_n$, where $Y_i = g^{-1}(X_i)$. Since $\widehat{F}_Y$ is a naive kernel distribution function, then it is guaranteed that $\sup_{y \in \mathbb{R}} |\widehat{F}_Y(y) - F_Y(y)| \to_{a.s.} 0$ [2], which implies

$$\sup_{x \in \Omega} |\widehat{F}_Y(g^{-1}(x)) - F_Y(g^{-1}(x))| \to_{a.s.} 0.$$

However, because $F_Y(g^{-1}(x)) = F_X(x)$, and it is clear that $\widehat{F}_Y(g^{-1}(x)) = \widetilde{F}_X(x)$, then this theorem is proven. $\square$

Even though it is not exactly related to the main topic of goodness-of-fit tests, it is worth to add that from $\widetilde{F}_X$ it is possible to derive another kernel-type estimator. It is clear that the density function f_X is equal to F'_X, then another boundary-free kernel density estimator is defined as $\widetilde{f}_X = \frac{d}{dx}\widetilde{F}_X$, which is

$$\widetilde{f}_X(x) = \frac{1}{nhg'(g^{-1}(x))} \sum_{i=1}^{n} K\left(\frac{g^{-1}(x) - g^{-1}(X_i)}{h}\right), \quad x \in \Omega. \qquad (5.9)$$

As $\widetilde{F}_X$ eliminates the boundary bias problem, this new estimator $\widetilde{f}_X$ does the same thing and can be a good competitor for other boundary bias reduction kernel density estimators. The bias and the variance of its are as follows.

Theorem 5.5 *Under condition D1–D5, also if g''' exists and f''_X is continuous, then the bias and the variance of $\widetilde{f}_X(x)$ are*

$$Bias[\widetilde{f}_X(x)] = \frac{h^2 c_2(x)}{2g'(g^{-1}(x))} \int_{-\infty}^{\infty} v^2 K(v)dv + o(h^2) \qquad (5.10)$$

$$Var[\widetilde{f}_X(x)] = \frac{f_X(x)}{nhg'(g^{-1}(x))} \int_{-\infty}^{\infty} K^2(v)dv + o\left(\frac{1}{nh}\right), \qquad (5.11)$$

where

$$c_2(x) = g'''(g^{-1}(x))f_X(x) + 3g''(g^{-1}(x))g'(g^{-1}(x))f'_X(x) + [g'(g^{-1}(x))]^3 f''_X(x).$$

Proof Using the similar reasoning as in the proof of Theorem 5.2,

$$E[\widehat{f}_X(x)] = \frac{1}{hg'(g^{-1}(x))} \int_{-\infty}^{\infty} K\left(\frac{g^{-1}(x) - y}{h}\right) f_Y(y)dy$$

$$= \frac{1}{g'(g^{-1}(x))} \int_{-\infty}^{\infty} f_Y(g^{-1}(x) - hv)K(v)dv$$

$$= \frac{f_Y(g^{-1}(x))}{g'(g^{-1}(x))} + \frac{h^2 f''_Y(g^{-1}(x))}{2g'(g^{-1}(x))} \int_{-\infty}^{\infty} v^2 K(v)dv + o(h^2),$$

and the bias formula is obtained. For the variance, first, calculate

$$\frac{1}{hg'(g^{-1}(x))} E\left[K^2\left(\frac{g^{-1}(x) - Y}{h}\right)\right] = \frac{1}{g'(g^{-1}(x))} \int_{-\infty}^{\infty} f_Y(g^{-1}(x) - hv)K^2(v)dv$$

$$= f_X(x) \int_{-\infty}^{\infty} K^2(v)dv + o(1),$$

and the rest is easily done. □

5.2.2 Boundary-Free Kernel-Smoothed KS and CvM Tests

As discussed before, the problem of the standard KS and CvM statistics is in the over-sensitivity near the center of distribution, because of the lack of smoothness of the empirical distribution function. Since the area around the center of distribution has the highest probability density, most of the realizations of the sample are there. As a result, $F_n(x)$ jumps a lot in those areas, and it causes some instability of estimation, especially when n is small. Conversely, though smoothing KS_n and CvM_n out using kernel distribution function can eliminate the oversensitivity near the center, the value of $\widehat{KS}$ and $\widehat{CvM}$ become larger than it should be when the data dealt with is supported on an interval smaller than the entire real line. This phenomenon is caused by the boundary problem.

Therefore, the clear solution to overcome the problems of standard and naive kernel goodness-of-fit tests together is to keep the smoothness of $\widehat{F}_X$ and to get rid of the boundary problem simultaneously. One of the ideas is by utilizing the boundary-free kernel distribution function estimator. Therefore, the proposed boundary-free kernel-smoothed Kolmogorov-Smirnov statistic is defined as

$$\widetilde{KS} = \sup_{x \in \mathbb{R}} |\widetilde{F}_X(x) - F(x)| \tag{5.12}$$

and boundary-free kernel-smoothed Cramér-von Mises statistic as

$$\widetilde{CvM} = n \int_{-\infty}^{\infty} [\widetilde{F}_X(x) - F(x)]^2 dF(x), \tag{5.13}$$

where $\widetilde{F}_X$ is the proposed estimator with a suitable function g.

Remark 5.2 Though the supremum and the integral are evaluated throughout the entire real line, computing them over Ω is sufficient, as $F_X(x) = \widetilde{F}_X(x)$ when $x \in \Omega^C$.

Although the formulas seem similar, one might expect both proposed tests are totally different from the standard KS and CvM tests. However, these two following theorems explain that the standard ones and the proposed methods turn out to be equivalent in the sense of distribution.

Theorem 5.6 *Let F_X and F be distribution functions on set Ω. If KS_n and $\widetilde{KS}$ are the standard and the proposed Kolmogorov-Smirnov statistics, respectively, then under the null hypothesis $F_X = F$,*

$$|KS_n - \widetilde{KS}| \rightarrow_p 0.$$

Proof First, consider the following inequality:

$$
\begin{aligned}
|KS_n - \widetilde{KS}| &= \left| \sup_{v \in \Omega} |F_n(v) - F(v)| - \sup_{z \in \Omega} |\widetilde{F}_X(z) - F(z)| \right| \\
&\leq \sup_{x \in \Omega} \left| |F_n(x) - F(x)| - |\widetilde{F}_X(x) - F(x)| \right| \\
&\leq \sup_{x \in \Omega} |F_n(x) - F(x) - \widetilde{F}_X(x) + F(x)| \\
&= \sup_{x \in \Omega} |\widetilde{F}_X(x) - F_n(x)|.
\end{aligned}
$$

Now, let $F_{n,Y}$ and $\widehat{F}_Y$ be the empirical distribution function and the naive kernel distribution function estimator, respectively, of $Y_1, Y_2, \ldots, Y_n$, where $Y_i = g^{-1}(X_i)$. Hence, it is guaranteed that $\sup_{y \in \mathbb{R}} |\widehat{F}_Y(y) - F_{n,Y}(y)| = o_p(n^{-1/2})$ [6], which further implies that

$$\sup_{x \in \Omega} |\widehat{F}_Y(g^{-1}(x)) - F_{n,Y}(g^{-1}(x))| \rightarrow_p 0$$

with rate $n^{-1/2}$. But, $\widehat{F}_Y(g^{-1}(x)) = \widetilde{F}_X(x)$ and $F_{n,Y}(g^{-1}(x)) = F_n(x)$, and the equivalency is proven. $\square$

Theorem 5.7 *Let F_X and F be distribution functions on set Ω. If CvM_n and $\widetilde{CvM}$ are the standard and the proposed Cramér-von Mises statistics, respectively, then under the null hypothesis $F_X = F$,*

$$|CvM_n - \widetilde{CvM}| \rightarrow_p 0.$$

Proof In this proof, the bandwidth is assumed to be $h = o(n^{-1/4})$. Define

$$\Delta_n = n \int_{-\infty}^{\infty} [\widetilde{F}_X(x) - F(x)]^2 dF(x) - n \int_{-\infty}^{\infty} [F_n(x) - F(x)]^2 dF(x).$$

Then,

$$\Delta_n = n \int_{-\infty}^{\infty} \left[\widetilde{F}_X(x) - F(x) - F_n(x) + F(x) \right]$$
$$\times \left[\widetilde{F}_X(x) - F(x) + F_n(x) - F(x) \right] dF(x)$$
$$= n \int_{-\infty}^{\infty} \frac{1}{n} \sum_{i=1}^{n} \left[W_i^*(x) - I_i^*(x) \right] \frac{1}{n} \sum_{j=1}^{n} \left[W_j^*(x) + I_j^*(x) \right] dF(x),$$

where

$$W_i^*(x) = W \left(\frac{g^{-1}(x) - g^{-1}(X_i)}{h} \right) - F(x) \quad \text{and} \quad I_i^*(x) = I(X_i \le x) - F(x).$$

Note that, if $i \ne j$, $W_i(\cdot)$ and $W_j(\cdot)$, also $I_i^*(\cdot)$ and $I_j^*(\cdot)$, are independent.

It follows from the Cauchy-Schwarz Inequality that $E(|\Delta_n|)$ is no greater than

$$n \int_{-\infty}^{\infty} \sqrt{ E \left[\left\{ \frac{1}{n} \sum_{i=1}^{n} (W_i^*(x) - I_i^*(x)) \right\}^2 \right] E \left[\left\{ \frac{1}{n} \sum_{j=1}^{n} (W_j^*(x) + I_j^*(x)) \right\}^2 \right] } \, dF(x).$$

Defining

$$b_n(x) = E \left[W \left(\frac{g^{-1}(x) - g^{-1}(X_i)}{h} \right) \right] - F(x) = O(h^2).$$

Hence, it follows from the independence that

$$E \left[\left\{ \frac{1}{n} \sum_{i=1}^{n} (W_i^*(x) - I_i^*(x)) \right\}^2 \right] = E \left[\left\{ \frac{1}{n} \sum_{i=1}^{n} (W_i^*(x) - b_n(x) - I_i^*(x)) \right\}^2 \right] + b_n^2(x)$$
$$= \frac{1}{n} E[\{W_1^*(x) - b_n(x) + I_1^*(x)\}^2] + b_n^2(x).$$

Furthermore,

$$E[\{W_1^*(x) - b_n(x) - I_1^*(x)\}^2]$$
$$= E[\{W_1^*(x) - I_1^*(x)\}^2] - 2b_n(x)E[W_1^*(x) - I_1^*(x)] + b_n^2(x)$$
$$= E[\{W_1^*(x)\}^2 - 2W_1^*(x)I_1^*(x) - \{I_1^*(x)\}^2] - b_n^2(x).$$

It follows from the mean squared error of $\widetilde{F}_X(x)$ that

$$E[\{W_1^*(x)\}^2] = F(x)[1 - F(x)] - 2hr_1 g'(g^{-1}(x)) f_X(x) + O(h^2).$$

From the definition, hence

$$E[W_1^*(x)I_1^*(x)]$$

$$= E\left[W\left(\frac{g^{-1}(x) - g^{-1}(X_1)}{h} \right) I(X_1 \le x) - F(x)W\left(\frac{g^{-1}(x) - g^{-1}(X_1)}{h} \right) \right.$$

$$\left. - F(x)I(X_1 \le x) + F^2(x) \right]$$

$$= E\left[W\left(\frac{g^{-1}(x) - g^{-1}(X_1)}{h} \right) I(X_1 \le x) \right] - F^2(x) - b_n(x)F(x).$$

For the first term,

$$E\left[W\left(\frac{g^{-1}(x) - Y}{h} \right) I(Y \le g^{-1}(x)) \right] = \int_{-\infty}^{g^{-1}(x)} W\left(\frac{g^{-1}(x) - y}{h} \right) f_Y(y)dy$$

$$= W(0)F(x) + F(x) \int_0^{\infty} K(v)dv + O(h).$$

Since $K(\cdot)$ is symmetric around the origin, then $W(0) = 1/2$. Thus,

$$E[W_1^*(x)I_1^*(x)] = F(x)[1 - F(x)] + O(h).$$

Next, $E\left[\{n^{-1} \sum_{i=1}^n (W_i^*(x) + I_i^*(x))\}^2 \right]$ will be evaluated. Using the bias term $b_n(x)$,

$$E\left[\left\{ \frac{1}{n} \sum_{i=1}^n (W_i^*(x) + I_i^*(x)) \right\}^2 \right] = E\left[\left\{ \frac{1}{n} \sum_{i=1}^n (W_i^*(x) - b_n(x) + I_i^*(x)) \right\}^2 \right] + b_n^2(x)$$

$$= \frac{1}{n} E[\{W_1^*(x) - b_n(x) + I_1^*(x)\}^2] + b_n^2.$$

Based on previous calculations,

$$E\left[\left\{ \frac{1}{n} \sum_{i=1}^n (W_i^*(x) + I_i^*(x)) \right\}^2 \right] = O\left(\frac{1}{n} + h^4 \right).$$

Therefore, if $h = o(n^{-1/4})$, then $E(|\Delta_n|) = o(1)$. Using the Markov Inequality, it is clear that $\Delta_n \to_p 0$, and then two statistics are equivalent under H_0. $\square$

Those equivalencies allow to use the same distribution tables of the standard goodness-of-fit tests for the new statistics. It means, with the same significance level α, the critical values are same.

5.2.3 Numerical Results

The results of some numerical studies will be shown in this subsection. The studies consist of two parts, the simulations of the proposed estimators $\widetilde{F}_X$ and $\widetilde{f}_X$, and then the results of the new goodness-of-fit tests $\widetilde{KS}$ and $\widetilde{CvM}$.

For the simulation to show the performances of the new distribution function estimator, the average integrated squared error (AISE) was calculated and repeated them 1000 times for each case. The naive kernel distribution function estimator $\widehat{F}_X$ and the proposed estimator $\widetilde{F}_X$ were compared. In the case of the proposed method, the two chosen mappings g^{-1} for each case. When $\Omega = \mathbb{R}^+$, the logarithm function $\log(x)$ and a composite of two functions $\Phi^{-1} \circ \gamma$ were chosen, where $\gamma(x) = 1 - e^x$. However, if $\Omega = [0, 1]$, probit and logit functions were utilized. With size 50, the generated samples were drawn from gamma $Gamma(2, 2)$, Weibull $Weibull(2, 2)$, standard log-normal $\log.N(0, 1)$, absolute-normal $abs.N(0, 1)$, standard uniform $U(0, 1)$, and beta distributions with three different sets of parameters ($Beta(1, 3)$, $Beta(2, 2)$, and $Beta(3, 1)$). The kernel function used here is the Gaussian Kernel and the bandwidths were chosen by cross-validation technique. Graphs of some chosen cases are shown as well in Fig. 5.1.

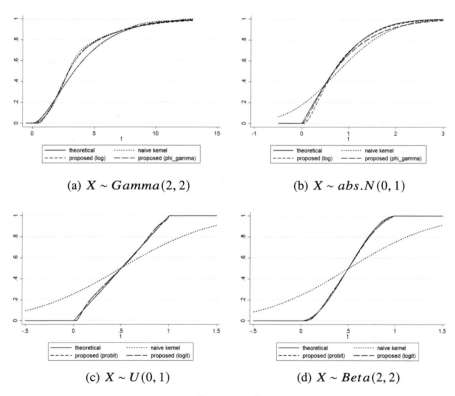

(a) $X \sim Gamma(2, 2)$ (b) $X \sim abs.N(0, 1)$

(c) $X \sim U(0, 1)$ (d) $X \sim Beta(2, 2)$

Fig. 5.1 Graphs comparisons of $F_X(x)$, $\widehat{F}_X(x)$, and $\widetilde{F}_X(x)$ for several distributions, with sample size $n = 50$

Table 5.1 AISE ($\times 10^5$) comparison of DF estimators

Distributions	$\widehat{F}_X$	$\widetilde{F}_{\log}$	$\widetilde{F}_{\Phi^{-1} \circ \gamma}$	$\widetilde{F}_{probit}$	$\widetilde{F}_{logit}$
$Gamma(2, 2)$	2469	2253	**2181**	–	–
$Weibull(2, 2)$	2224	**1003**	1350	–	–
$\log .N(0, 1)$	1784	1264	**1254**	–	–
$abs.N(0, 1)$	2517	**544**	727	–	–
$U(0, 1)$	5074	–	–	**246**	248
$Beta(1, 3)$	7810	–	–	**170**	172
$Beta(2, 2)$	6746	–	–	**185**	188
$Beta(3, 1)$	7801	–	–	**154**	156

In Table 5.1, the proposed estimator outperformed the naive kernel distribution function. Though the differences are not so big in the cases of gamma, Weibull, and the log-normal distributions, the gaps are glaring in the absolute-normal case or when the support of the distributions is the unit interval. The cause of this phenomenon might be seen in Fig. 5.1.

Albeit the shapes of $\widetilde{F}_{\log}$ and $\widetilde{F}_{\Phi^{-1} \circ \gamma}$ are more similar to the theoretical distribution in Fig. 5.1a, but it has to be admitted that the shape of $\widehat{F}_X$ is not so much different with the rests. However, in Fig. 5.1b, c, and d, it is obvious that the naive kernel distribution function is too far-off the mark, particularly in the case of $\Omega = [0, 1]$. As the absolute-normal, uniform, and beta distributions have quite high probability density near the boundary point $x = 0$ (also $x = 1$ for unit interval case), the naive kernel estimator spreads this "high density information" around the boundary regions. However, since $\widehat{F}_X$ cannot detect the boundaries, it puts this "high density information" outside the support as well. This is not happening too severely in the case of Fig. 5.1a because the probability density near $x = 0$ is fairly low. Hence, although the value of $\widehat{F}_X(x)$ might be still positive when $x \approx 0^-$, it is not so far from 0 and vanishes quickly

Remark 5.3 Figure 5.1c and d also gave a red alert if the naive kernel distribution function is used in place of empirical distribution for goodness-of-fit tests. As the shapes of $\widehat{F}_X$ in Fig. 5.1c and d resemble the normal distribution function a lot, if $H_0 : X \sim N(\mu, \sigma^2)$ is tested, the tests may not reject the null hypothesis. This shall cause the increment of type-2 error.

Remark 5.4 It is worth to note that in Table 5.1, even though $\widetilde{F}_{probit}$ performed better, its differences are too little to claim that it outperformed $\widetilde{F}_{logit}$. From here, it can be concluded that probit and logit functions work pretty much the same for $\widetilde{F}_X$.

Since $\widehat{f}_X$ was also introduced as a new boundary-free kernel density estimator, some illustrations of its performances in this subsection were also provided. Under the same settings as in the simulation study of the distribution function case, the results of its simulation can be seen in Table 5.2 and Fig. 5.2.

From AISE point of view, once again the proposed estimator outperformed the naive kernel one, and huge gaps happened as well when the support of the distribution

Table 5.2 AISE ($\times 10^5$) comparison of density estimators

Distributions	$\widehat{f}_X$	$\widetilde{f}_{\log}$	$\widetilde{f}_{\Phi^{-1}\circ\gamma}$	$\widetilde{f}_{probit}$	$\widetilde{f}_{logit}$
$Gamma(2, 2)$	925	744	**624**	–	–
$Weibull(2, 2)$	6616	**3799**	3986	–	–
$\log.N(0, 1)$	7416	3569	**2638**	–	–
$abs.N(0, 1)$	48005	34496	**14563**	–	–
$U(0, 1)$	36945	–	–	**14235**	21325
$Beta(1, 3)$	109991	–	–	**18199**	28179
$Beta(2, 2)$	52525	–	–	**5514**	6052
$Beta(3, 1)$	109999	–	–	**17353**	28935

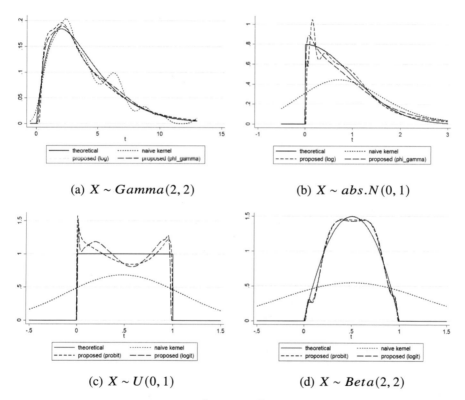

(a) $X \sim Gamma(2, 2)$ (b) $X \sim abs.N(0, 1)$

(c) $X \sim U(0, 1)$ (d) $X \sim Beta(2, 2)$

Fig. 5.2 Graphs comparisons of $f_X(x)$, $\widehat{f}_X(x)$, and $\widetilde{f}_X(x)$ for several distributions, with sample size $n = 50$

is the unit interval. Some interests might be arisen from Fig. 5.2b, c, and d, as the graphs of $\widehat{F}_X$ are too different from the theoretical ones, and more similar to the Gaussian bell shapes instead.

Next, the results of the simulation studies regarding the new Kolmogorov-Smirnov and Cramér-von Mises tests in this part are provided. As a measure of comparison,

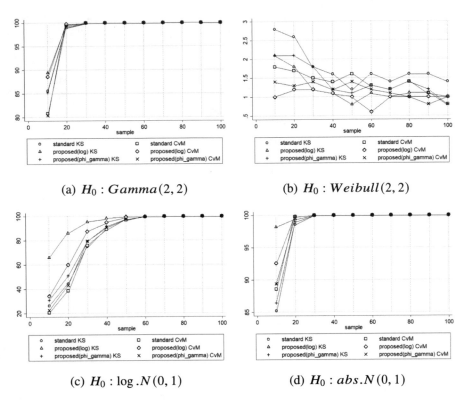

(a) $H_0 : Gamma(2, 2)$

(b) $H_0 : Weibull(2, 2)$

(c) $H_0 : \log .N(0, 1)$

(d) $H_0 : abs.N(0, 1)$

Fig. 5.3 Simulated percentage (%) of rejecting null hypothesis when the samples were drawn from $Weibull(2, 2)$

the percentage of rejecting several null hypotheses when the samples were drawn from certain distributions was calculated. When the actual distribution and the null hypothesis are same, the percentage should be close to $100\alpha\%$ (significance level in percent). However, if the real distribution does not match the H_0, the percentage is expected to be as large as possible. To illustrate how the behaviors of the statistics change, a sequential number of sample sizes were generated, started from 10 until 100, with 1000 repetitions for each case. The chosen level of significance is $\alpha = 0.01$, and the standard tests were compared with the proposed tests.

From Fig. 5.3, it can be seen that the modified KS and CvM tests outperformed the standard ones, especially the proposed KS test with logarithm as the bijective transformation. From Fig. 5.3a, c, and d, KS test with logarithm function has the highest percentage of rejecting H_0 even when the sample sizes were still 10. However, even though the new CvM test with logarithmic function was always the second highest in the beginning, $\widetilde{CvM}_{\log}$ was also the first one that reached 100%. On the other hand, based on Fig. 5.3b, all statistical tests (standard and proposed) were having similar stable behaviors, as their numbers were still in the interval $0.5-2\%$. However

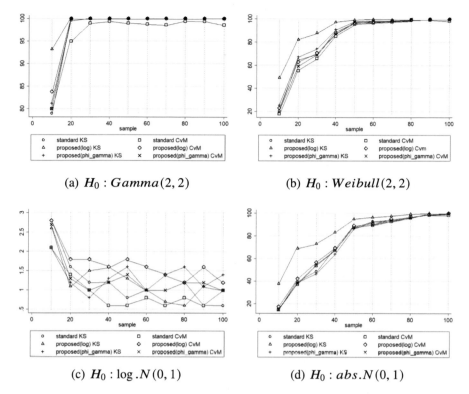

(a) $H_0 : Gamma(2, 2)$

(b) $H_0 : Weibull(2, 2)$

(c) $H_0 : \log .N(0, 1)$

(d) $H_0 : abs.N(0, 1)$

Fig. 5.4 Simulated percentage (%) of rejecting null hypothesis when the samples were drawn from $\log .N(0, 1)$

at this time, $\widetilde{CvM}_{\log}$ performed slightly better than others, because its numbers in general were the closest to 1%.

Similar things happened when the samples were drawn from the standard lognormal distribution, which the proposed methods outperformed the standard ones. However, this time, the modified KS test with $g^{-1} = \log$ always gave the best results. Yet, some notes may be taken from Fig. 5.4. First, although when $n = 10$ all the percentages were far from 1% in Fig. 5.4c, but, after $n = 20$, every test went stable inside 0.5−2% interval. Second, as seen in Fig. 5.4d, it seems difficult to reject $H_0 : abs.N(0, 1)$ when the actual distribution is $\log .N(0, 1)$, even $\widetilde{KS}_{\log}$ could only reach 100% rejection after $n = 80$. While, on the other hand, it was quite easy to reject $H_0 : Gamma(2, 2)$ as most of the tests already reached 100% rejection when $n = 20$.

Something more extreme happened in Fig. 5.5, as all of the tests could reach 100% rejection rate since $n = 30$, even since $n = 10$ in Fig. 5.5d. Though seems strange, the cause of this phenomenon is obvious. The shape of the distribution function of $Beta(1, 3)$ is so different from other three distributions in this study, especially with $Beta(3, 1)$. Hence, even with a small sample size, the tests could reject the false

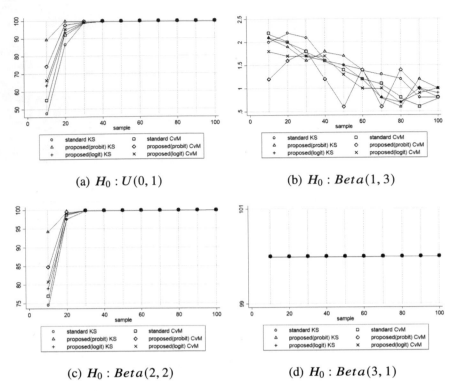

(a) $H_0 : U(0, 1)$

(b) $H_0 : Beta(1, 3)$

(c) $H_0 : Beta(2, 2)$

(d) $H_0 : Beta(3, 1)$

Fig. 5.5 Simulated percentage (%) of rejecting null hypothesis when the samples were drawn from $Beta(1, 3)$

null hypothesis. However, still the proposed tests worked better than the standard goodness-of-fit tests, because before all the tests reached 100% point, the standard KS and CvM tests had the lowest percentages.

From these numerical studies, it can be concluded that both the standard and the proposed KS and CvM tests will give the same result when the sample size is large. However, if the sample size is small, the proposed methods will give better and more reliable results.

In this analysis, the UIS Drug Treatment Study Data [39] is used again to show the performances of the proposed methods for real data. The dataset records the result of an experiment about how long someone who got drug treatment to relapse the drug again. With a total of 623 observations, the variable used in the calculation was the "age" variable, which represents the age of the individual when they were admitted to drug rehabilitation for the first time.

For simplicity, gamma, log-normal, and Weibull distributions were taken for null hypothesis of the tests. Hence, using maximum likelihood estimation, $\Gamma(27.98, 1.16)$, $\log.N(3.46, 0.04)$, and $Weibull(5.55, 34.95)$ were taken as candidates.

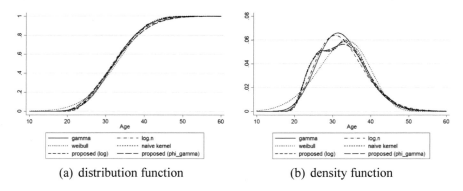

(a) distribution function (b) density function

Fig. 5.6 Comparisons of $Gamma(27.98, 1.16)$, $\log.N(3.46, 0.04)$, $Weibull(5.55, 34.95)$, naive kernel estimator, and the proposed method ($\log$ and $\Phi^{-1} \circ \gamma$) for UIS data with $n = 623$

Table 5.3 The p-values (%) from several tests for UIS data ($n = 623$)

Test \ H_0	$Gamma$	$\log.N$	$Weibull$
KS_n	2.77	2.69	**18.20**
$\widetilde{KS}$	3.39	3.35	**14.24**
$\widetilde{KS}_{\log}$	2.69	2.66	**19.14**
$\widetilde{KS}_{\Phi^{-1}\circ\gamma}$	2.79	2.74	**17.35**
CvM_n	2.23	2.25	**15.25**
$\widetilde{CvM}$	3.64	3.54	**11.91**
$\widetilde{CvM}_{\log}$	2.18	2.13	**17.05**
$\widetilde{CvM}_{\Phi^{-1}\circ\gamma}$	2.26	2.19	**13.69**

It is difficult to infer anything from Fig. 5.6a, but it can be seen from Fig. 5.6b that the shapes of the three estimators resemble the density of Weibull distribution the most, compared to the other two distributions. That statement was proven in table 5.3 that all the tests accepted $H_0 : Weibull(5.55, 34.95)$ and rejected the others. Hence, it can be concluded that the variable "age" from UIS data is under Weibull distribution.

Though Fig. 5.6 and Table 5.3 show good results, the performances of the different tests cannot be really compared because the conclusions are same. It is obvious that due to the large number of observations, all estimators converged to the true distribution of the data, resulting to all tests performed similarly. Hence, to find the difference among the results of the tests, 50 observations from UIS data were drawn again to create "new" sample, and did the same tests and analysis as before.

Now, after reducing the sample size to 50, it is clear that something different in Table 5.4, which is that the naive kernel KS and CvM tests no longer rejected gamma hypothesis. Also, from Fig. 5.7, it is glaring that the shapes of the graphs of $\widehat{F}_X$ and $\widehat{f}_X$ are, although still close to Weibull, shifting to the shape of gamma distribution and density. Not only that, the standard KS and CvM tests also *almost* accepted gamma

Table 5.4 The p-values (%) from several tests for UIS data ($n = 50$)

Test \ H_0	Gamma	$\log.N$	Weibull
KS_n	4.76	4.18	**18.13**
$\widehat{KS}$	**5.14**	4.89	**14.81**
$\widetilde{KS}_{\log}$	4.08	3.92	**19.05**
$\widetilde{KS}_{\Phi^{-1}\circ\gamma}$	4.18	3.97	**17.89**
CvM_n	4.22	4.14	**14.23**
$\widehat{CvM}$	**5.23**	4.93	**12.19**
$\widetilde{CvM}_{\log}$	4.01	3.86	**17.14**
$\widetilde{CvM}_{\Phi^{-1}\circ\gamma}$	4.27	4.12	**15.35**

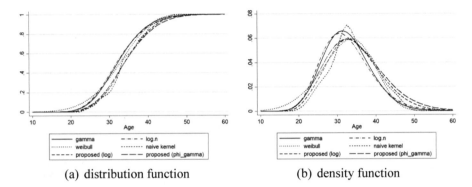

    (a) distribution function   (b) density function

Fig. 5.7 Comparisons of $Gamma(27.98, 1.16)$, $\log.N(3.46, 0.04)$, $Weibull(5.55, 34.95)$, naive kernel estimator, and the proposed method (log and $\Phi^{-1}\circ\gamma$) for UIS data with $n = 50$

hypothesis as their p-values are close to 5%, leaving the proposed methods stayed with the same conclusions as when the whole 623 observations were being analyzed. From this trend, it is safe to predict that, if the sample size is reduced further, the standard and naive kernel tests would accept the wrong null hypothesis, resulting to the increment of type-2 error.

From this real data analysis, it can be concluded that the proposed tests and estimators are more stable, even when the sample size is small. Oppositely, the standard and naive kernel KS and CvM tests, though work really well when dealing with a large number of observations, their results are not really reliable if the sample size is not large enough.

5.3 Smoothed Nonparametric Tests and Approximation of p-Value

Let $X_1, X_2, \ldots, X_n$ be independently and identically distributed (*i.i.d.*) random variables with a distribution function $F(x - \theta)$, where the associated density function satisfies $f(-x) = f(x)$ and θ is an unknown location parameter. Here, we consider a test and the confidence interval of the parameter θ. This setting is called a one-sample location problem. Numerous nonparametric test statistics have been proposed to test the null hypothesis $H_0 : \theta = 0$ vs. $H_1 : \theta > 0$, e.g., the sign test and Wilcoxon's signed rank test. These tests are distribution-free and have discrete distributions. As pointed out by [7], because of the discreteness of the test statistics, the p-values jump in response to a small change in data values when the sample size n is small or moderate. A smoothed version of the sign test and then obtained an Edgeworth expansion is discussed in [7]. In particular, they introduced a smoothed median estimator and a corresponding smoothed sign test. The test was, however, not distribution-free. Their smoothed sign test has good properties, but Pitman's asymptotic relative efficiency (*A.R.E.*) did not coincide with that of the ordinary sign test. Furthermore, to use the Edgeworth expansion, they needed estimators of the unknown parameters.

In this section, we first consider another smoothed sign test that is based on a kernel estimator of the distribution function and examine its asymptotic properties. We show that the difference between the two sign tests converges to zero in probability. In addition, we obtain an Edgeworth expansion, being free of the underlying distribution. Next, we discuss a smoothed Wilcoxon's signed rank test. We show that the difference between the two Wilcoxon's tests converges to zero in probability as well and study an Edgeworth expansion.

Let us define the indicator function $I(A) = 1$ (if A occurs), $= 0$ (if A fails); the sign test is equivalent to

$$S = S(\mathbf{X}) = \sum_{i=1}^{n} I(X_i \geq 0),$$

where $\mathbf{X} = (X_1, X_2, \ldots, X_n)^T$. Wilcoxon's signed rank test is equivalent to the Mann-Whitney test:

$$W = W(\mathbf{X}) = \sum_{1 \leq i \leq j \leq n} I(X_i + X_j \geq 0).$$

Now, put $s = S(\mathbf{x})$ and $w = W(\mathbf{x})$ for observed values $\mathbf{x} = (x_1, x_2, \ldots, x_n)^T$. If the p-value $P_0(S \geq s)$ ($P_0(W \geq w)$), where $P_0(\cdot)$ denotes a probability under the null hypothesis H_0, is small enough, we reject the null hypothesis H_0.

Moreover, let us define

$$\Omega_{|x|} = \left\{ \mathbf{x} \in \mathbb{R}^n \mid |x_1| < |x_2| < \cdots < |x_n| \right\}$$

Table 5.5 Number of samples in which S and W have comparatively smaller p-values

	Sample size	$n = 10$	$n = 20$	$n = 30$
$z_{0.90}$	S	25	69080	59092679
	W	82	94442	87288529
	W/S	3.28	1.367	1.477
$z_{0.95}$	S	25	32705	30857108
	W	48	47387	43957510
	W/S	1.92	1.449	1.425
$z_{0.975}$	S	5	12704	14028374
	W	21	21267	22049240
	W/S	4.2	1.674	1.572

and

$$\Omega_\alpha = \left\{ \mathbf{x} \in \Omega_{|x|} \; \left\| \; \frac{s - E_0(S)}{\sqrt{V_0(S)}} \geq z_{1-\alpha}, \quad \text{or} \quad \frac{w - E_0(W)}{\sqrt{V_0(W)}} \geq z_{1-\alpha} \right. \right\},$$

where $z_{1-\alpha}$ is the $(1 - \alpha)th$ quantile of the standard normal distribution $N(0, 1)$, and $E_0(\cdot)$ and $V_0(\cdot)$ are, respectively, the expectation and variance under H_0. The observed values $S(\mathbf{x})$ and $W(\mathbf{x})$ are invariant under a permutation of $x_1, \ldots, x_n$, so it is sufficient to consider the case that $|x_1| < |x_2| < \cdots < |x_n|$; there are 2^n times combinations of $sign(x_i) = \pm 1 (i = 1, \ldots, n)$. We count samples in which the exact p-value of one test is smaller than the p-value of the other test in the tail area Ω_α. Table 5.5 shows the number of samples in which the p-value of S (W) is smaller than that of W (S) in the tail area. In Table 5.5, row S indicates a number of samples in which the p-value of S is smaller than that of W, row W means the number of samples in which the p-value of W is smaller than that of S, and W/S is the ratio of W and S. For each sample, there is one tie of p-values.

Remark 5.5 Table 5.5 shows that W is preferable if one wants a small p-value and that S is preferable if one does not want to reject the null hypothesis H_0. Thus, a practitioner could make an arbitrary choice of the test statistics. This problem comes from the discreteness of the distributions of the test statistics.

On the other hand, it is possible to use an estimator of $F(0)$ as a test statistic. Define the empirical distribution function by

$$F_n(x) = \frac{1}{n} \sum_{i=1}^{n} I(X_i \leq x).$$

Then, $F_n(0)$ is equivalent to the sign test S, that is,

$$S = n - nF_n(0-).$$

As usual, a kernel estimator $\widetilde{F}_n$ can be used to get a smooth estimator of the distribution function. It is natural to use $\widetilde{F}_n(0)$ as a smoothed sign test. Let k be a kernel function that satisfies

$$\int_{-\infty}^{\infty} k(u)du = 1,$$

and write K be the integral of k,

$$K(t) = \int_{-\infty}^{t} k(u)du.$$

The kernel estimator of $F(x)$ is defined by

$$\widetilde{F}_n(x) = \frac{1}{n} \sum_{i=1}^{n} K\left(\frac{x - X_i}{h_n}\right),$$

where h_n is a bandwidth that satisfies $h_n \to 0$ and $nh_n \to \infty$ ($n \to \infty$). We can use

$$\widetilde{S} = n - n\widetilde{F}_n(0) = n - \sum_{i=1}^{n} K\left(-\frac{X_i}{h_n}\right)$$

for testing H_0 and can regard $\widetilde{S}$ as the smoothed sign test. Under H_0, the main terms of the asymptotic expectation and variance of $\widetilde{S}$ do not depend on F; i.e., they are asymptotically distribution-free. Furthermore, we can obtain an Edgeworth expansion, being free of F.

We can construct a smoothed Wilcoxon's signed rank test in a similar fashion. Since the main term of the Mann-Whitney statistic can be regarded as an estimator of the probability $P\left(\frac{X_1 + X_2}{2} > 0\right)$, we propose the following smoothed test statistic:

$$\widetilde{W} = \frac{n(n + 1)}{2} - \sum_{1 \leq i \leq j \leq n} K\left(-\frac{X_i + X_j}{2h_n}\right).$$

The smoothed test $\widetilde{W}$ is not distribution-free. However, under H_0, the asymptotic expectation and variance do not depend on F, and we can obtain the Edgeworth expansion of $\widetilde{W}$. The resulting Edgeworth expansion does not depend on F if we use a symmetric fourth-order kernel and bandwidth of $h_n = o(n^{-1/4})$.

Also, we will show that the difference between the standardized S and $\widetilde{S}$ and the difference between the standardized W and $\widetilde{W}$ go to zero in probability. Accordingly, the smoothed test statistics are equivalent in the sense of the first-order asymptotic.

5.3.1 Asymptotic Properties of Smoothed Tests

We assume that the kernel k is symmetric, i.e., $k(-u) = k(u)$. Using the properties of the kernel estimator, we can obtain expectations $E_\theta(\widetilde{S})$, $E_\theta(\widetilde{W})$ and variances $V_\theta(\widetilde{S})$, $V_\theta(\widetilde{W})$. Because of the symmetry of the underlying distribution f and the kernel k, we get

$$F(-x) = 1 - F(x) \quad \text{and} \quad \int_{-\infty}^{\infty} uk(u)du = 0.$$

Let us define

$$e_1(\theta) = E_\theta \left[1 - K \left(-\frac{X_1}{h_n} \right) \right].$$

Using the transformation $u = -x/h_n$, integration by parts, and a Taylor-series expansion, we get

$$e_1(\theta) = 1 - \int_{-\infty}^{\infty} K(u) f(-\theta - h_n u) \frac{1}{h_n} du$$

$$= 1 - \int_{-\infty}^{\infty} k(u) F(-\theta - h_n u) du$$

$$= F(\theta) + O(h_n^2),$$

which yields

$$E_\theta(\widetilde{S}) = n \left\{ F(\theta) + O(h_n^2) \right\}.$$

Similarly, we have

$$E_\theta \left[K^2 \left(-\frac{X_1}{h_n} \right) \right] = F(-\theta) + O(h_n),$$

hence,

$$V_\theta(\widetilde{S}) = n \left[\{1 - F(\theta)\} F(\theta) + O(h_n) \right].$$

On the other hand, since $\widetilde{W}$ takes the form of the U-statistic, we can use asymptotic properties of the U-statistic. The expectation and variance of $\widetilde{W}$ are given by

$$E_\theta(\widetilde{W}) = \frac{n(n+1)}{2} \left\{ G(\theta) + O(h_n^2) \right\},$$

$$V_\theta(\widetilde{W}) = n(n+1)^2 \left\{ \int_{-\infty}^{\infty} F^2(u + 2\theta) f(u) du - G^2(\theta) + O(h_n^2) \right\},$$

where

$$G(\theta) = \int_{-\infty}^{\infty} F(2\theta + u) f(u) du$$

is the distribution function of $(X_1 + X_2)/2$.

Direct computations yield the following theorem.

Theorem 5.8 *Let us assume that f' exists and is continuous in a neighborhood of $-\theta$, and $h_n = cn^{-d} (c > 0, \frac{1}{4} < d < \frac{1}{2})$. If*

$$0 < \lim_{n \to \infty} V_\theta \left[1 - K\left(-\frac{X_1}{h_n} \right) \right] < \infty,$$

$$0 < \lim_{n \to \infty} Cov_\theta \left[1 - K\left(-\frac{X_1 + X_2}{2h_n} \right), 1 - K\left(-\frac{X_1 + X_3}{2h_n} \right) \right] < \infty,$$

and the kernel k is symmetric around zero, then,

$$\lim_{n \to \infty} E_\theta \left\{ \frac{S - E_\theta(S)}{\sqrt{V_\theta(S)}} - \frac{\widetilde{S} - E_\theta(\widetilde{S})}{\sqrt{V_\theta(\widetilde{S})}} \right\}^2 = 0,$$

$$\lim_{n \to \infty} E_\theta \left\{ \frac{W - E_\theta(W)}{\sqrt{V_\theta(W)}} - \frac{\widetilde{W} - E_\theta(\widetilde{W})}{\sqrt{V_\theta(\widetilde{W})}} \right\}^2 = 0.$$

Proof The complete proof can be found in [8]. □

Since S and W are asymptotically normal, $\widetilde{S}$ and $\widetilde{W}$ are also asymptotically normal. Pitman's *A.R.E.*s of $\widetilde{S}$ and $\widetilde{W}$ coincide with S and W, respectively.

For the sign test S, it is difficult to improve the normal approximation because of the discreteness of the distribution function of S. The standardized sign test S takes values with jump order $n^{-1/2}$, so we cannot prove the validity of the formal Edgeworth expansion. On the other hand, since $\widetilde{S}$ is a smoothed statistic and has a continuous type distribution, we can obtain an Edgeworth expansion and prove its validity. The Edgeworth expansion is discussed and proved its validity for the kernel estimators in [9]. An explicit formula when $h_n = cn^{-d} (c > 0, \frac{1}{4} < d < \frac{1}{2})$ is obtained in [10]. The paper [11] proved the validity of the Edgeworth expansion of the U-statistic with an $o(n^{-1})$ residual term. Since the standardized W and $\widetilde{W}$ are asymptotically equivalent, we can obtain the Edgeworth expansion of $\widetilde{W}$. The resulting Edgeworth approximations do not depend on the underlying distribution F, if we use the fourth-order kernel, i.e.,

$$\int u^\ell k(u) du = 0 \ (\ell = 1, 2, 3) \quad \text{and} \quad \int u^4 k(u) du \neq 0.$$

Using the results of [9] and [10], we can prove the following theorem.

Theorem 5.9 *Let us assume that the conditions of Theorem 5.8 hold and the kernel k is symmetric. If $|f'(x)| \le M (M > 0)$, $\int |u^4 k(u)| du < \infty$ and the bandwidth satisfies $h_n = cn^{-d}$ $(c > 0, \frac{1}{4} < d < \frac{1}{2})$, then*

$$P_0 \left(\frac{\tilde{S} - E_0(\tilde{S})}{\sqrt{V_0(\tilde{S})}} \le y \right) = \Phi(y) - \frac{1}{24n}(y^3 - 3y)\phi(y) + o(n^{-1}),$$

$$P_0 \left(\frac{\tilde{W} - E_0(\tilde{W})}{\sqrt{V_0(\tilde{W})}} \le y \right) = \Phi(y) - \left(\frac{7}{20}y^3 - \frac{21}{20}y \right)\phi(y) + o(n^{-1}).$$

Proof The complete proof can be found in [8]. □

The Edgeworth expansions in Theorem 5.9 do not depend on the underlying distribution F. However, in order to use the normal approximations or the Edgeworth expansions, we have to obtain approximations of $E_0(\tilde{S})$, $V_0(\tilde{S})$, $E_0(\tilde{W})$ and $V_0(\tilde{W})$. Let us define

$$A_{i,j} = \int_{-\infty}^{\infty} K^i(u)k(u)u^j du.$$

We have the following higher order approximations of the expectations and variances under the null hypothesis H_0.

Theorem 5.10 *Let us assume that the kernel is symmetric, and let M_1, M_2, and M_3 be positive constants. If exactly one of the following conditions holds: (a) $|f^{(5)}(x)| \le M_1$ and $h_n = o(n^{-1/4})$, (b) $|f^{(4)}(x)| \le M_2$ and $h_n = o(n^{-3/10})$, (c) $|f^{(3)}(x)| \le M_3$ and $h_n = o(n^{-1/3})$, then,*

$$E_0(\tilde{S}) = \frac{n}{2} + o(n^{-1/2}),$$

$$V_0(\tilde{S}) = \frac{n}{4} - 2nh_n f(0)A_{1,1} - \frac{nh_n^3}{3} f''(0)A_{1,3} + o(1),$$

$$E_0(\tilde{W}) = \frac{n(n+1)}{4} + o(n^{1/2}), \tag{5.14}$$

$$V_0(\tilde{W}) = \frac{n^2(2n+3)}{24} - 4n^3 h_n^2 A_{0,2} \int_{-\infty}^{\infty} \{f(x)\}^3 dx + o(n^2). \tag{5.15}$$

Proof The complete proof can be found in [8]. □

Remark 5.6 In order to get the above approximations, we used a Taylor-series expansion of the integral. We can divide up the integral at discrete points, so we do not need to worry about the differentiability of the density function at finite number of points.

5.3.2   Selection of Bandwidth and Kernel Function

We discuss the selection of the bandwidth and the kernel function. Azzalini in [9] recommended a bandwidth of $cn^{-1/3}$ for the estimation of the distribution function. Actually, we compared several bandwidths in simulation studies and found that the approximations were not good when the convergence rate of the bandwidth was slower than $n^{-1/3}$. When the convergence rate of the bandwidth was faster than $n^{-1/3}$, the approximations were similar to the case of $n^{-1/3}$. Thus, hereafter, we will use the bandwidth $h_n = n^{-1/3}$. The well-known Epanechnikov Kernel

$$k_{e,2}(u) = \frac{3}{4}(1 - u^2)I(|u| \le 1)$$

is said to be the optimal kernel function for density estimation ($k_{e,2}$ is a second-order kernel). We observed from several simulation studies that the second-order kernels ($A_{0,1} = 0$, $A_{0,2} \ne 0$) do not give good approximations of the p-values. We compare normal approximations of $\widetilde{S}$ based on $k_{e,2}(u)$ and a modified fourth-order kernel

$$k_{e,4}(u) = \frac{15}{8}\left(1 - \frac{7}{3}u^2\right)k_{e,2}(u)$$

with the bandwidth $h_n - n^{-1/3}$. We simulated the following probabilities from 100,000 random samples from normal, logistic, and double exponential distributions:

$$P_0\left(\frac{\widetilde{S} - n/2}{\sqrt{n/4}} \ge z_{1-\alpha}\right).$$

Table 5.6 shows that the fourth-order kernel gives good approximations. Note that in the table, the normal distribution is denoted as $N(0, 1)$, the logistic distribution as Logis., and the double exponential distribution as D.Exp. We performed similar simulations using a Gaussian kernel and obtained similar results. The choice of kernel function does not affect the approximations of the p-values, but the order degree of the kernel is important, as expected. Although the fourth-order kernels lose the monotonicity of the distribution functions of $\widetilde{S}$ and $\widetilde{W}$, they lose monotonicity at most 3% points of x when $n = 10$.

Remark 5.7 If we use the normal approximations of the standardized $\widetilde{S}$ and $\widetilde{W}$, $A_{1,1}$ affects the approximations. We recommend the fourth-order kernel because the value of $A_{1,1}$ with the fourth-order kernel is much smaller than that of the second-order one.

Since the distributions of $\widetilde{S}$ and $\widetilde{W}$ depend on F, we compared the significance probabilities of $\widetilde{S}$ and $\widetilde{W}$ in simulations. We used the kernel $k_{e,4}$ and bandwidth $h_n = n^{-1/3}$. We estimated the significance probabilities in the tail area $\widetilde{\Omega}$ from 100,000 random samples from a normal distribution:

Table 5.6 p-value approximations of smoothed sign $\widetilde{S}$ with $k_{e,2}$ and $k_{e,4}$ ($h_n = n^{-1/3}$)

$\alpha = 0.05$		$n = 10$	$n = 20$	$n = 30$
$k_{e,2}$	N(0,1)	0.03437	0.03803	0.03834
$k_{e,4}$	N(0,1)	0.05224	0.05397	0.05374
$k_{e,2}$	Logis.	0.04038	0.04256	0.04293
$k_{e,4}$	Logis.	0.05382	0.05395	0.05347
$k_{e,2}$	D.exp	0.02603	0.03062	0.03193
$k_{e,4}$	D.exp	0.04272	0.04468	0.04797

Table 5.7 Number of samples in which S and W have comparatively smaller p-values ($k_{e,4}$, $h_n = n^{-1/3}$)

	Sample size	$n = 10$	$n = 20$	$n = 30$
$z_{0.90}$	$\widetilde{S}$	5658	5978	6142
	$\widetilde{W}$	7263	7066	7050
	$\widetilde{W}/\widetilde{S}$	1.284	1.182	1.148
$z_{0.95}$	$\widetilde{S}$	2921	3017	3164
	$\widetilde{W}$	3407	3616	3515
	$\widetilde{W}/\widetilde{S}$	1.166	1.199	1.111
$z_{0.975}$	$\widetilde{S}$	1133	1440	1599
	$\widetilde{W}$	1628	1785	1780
	$\widetilde{W}/\widetilde{S}$	1.437	1.240	1.113

$$\widetilde{\Omega}_\alpha = \left\{ \mathbf{x} \in \mathbb{R}^n \ \left\| \ \frac{\widetilde{s}(\mathbf{x}) - E_0(\widetilde{S})}{\sqrt{V_0(\widetilde{S})}} \geq z_{1-\alpha}, \quad \text{or} \quad \frac{\widetilde{w}(\mathbf{x}) - E_0(\widetilde{W})}{\sqrt{V_0(\widetilde{W})}} \geq z_{1-\alpha} \right. \right\}.$$

For the simulated sample $\mathbf{x} \in \mathbb{R}^n$, we calculated the p-values based on the normal approximation. In Table 5.7, $\widetilde{S}$ means that the p-values of $\widetilde{S}$ are smaller than that of $\widetilde{W}$, etc. Comparing Tables 5.5 and 5.7, we can see that the differences between the p-values of $\widetilde{S}$ and $\widetilde{W}$ are smaller than those of S and W.

Next, we checked how close the ordinary and smoothed tests are to each other when the sample size n is small. Since S has a discrete distribution, we chose a nearest value α' to 0.05, i.e., $P(S \geq s_{\alpha'}) = \alpha' \approx 0.05$. After that, we simulated the p-values $P(S \geq s_{\alpha'})$ and $P(\widetilde{S} \geq \frac{n}{2} + \frac{\sqrt{n}}{2} z_{1-\alpha'})$ from 100,000 repetitions for underlying normal ($N(0, 1)$) and double exponential (D.exp) distributions.

Table 5.8 shows that the difference between the smoothed and ordinary sign tests is small, so we can regard $\widetilde{S}$ as a smoothing statistic of S. We got similar results for $\widetilde{W}$ and W.

Table 5.8 Closeness of p-values of S & $\widetilde{S}$, and W & $\widetilde{W}$ ($k_{e,4}$, $h_n = n^{-1/3}$)

$\alpha = 0.05$		$n = 10$	$n = 20$	$n = 30$		$n = 10$	$n = 20$	$n = 30$
S	N(0,1)	0.05474	0.05867	0.04950	D.exp	0.05446	0.05617	0.04947
$\widetilde{S}$	N(0,1)	0.04981	0.05358	0.04548	D.exp	0.04434	0.04915	0.04187
W	N(0,1)	0.05271	0.04797	0.05092	D.exp	0.05271	0.04942	0.04998
$\widetilde{W}$	N(0,1)	0.05226	0.04864	0.05046	D.exp	0.05117	0.04879	0.04962

5.3.3 Higher Order Approximation

We discuss higher order approximations based on Edgeworth expansions. If the conditions of $A_{1,1} = 0$ and $A_{1,3} = 0$ hold, we can use the Edgeworth expansion of $\widetilde{S}$. If the kernel is fourth-order symmetric, $A_{0,2} = 0$ and we can use the Edgeworth expansion of $\widetilde{W}$. The conditions of $A_{1,1} = 0$ and $A_{1,3} = 0$ seem restrictive, but we can still construct the desired kernel. Let us define

$$k^*(u) = \left(\frac{1}{4}(\sqrt{105} - 3) + \frac{1}{2}(5 - \sqrt{105})|u| \right) I(|u| \leq 1),$$

which is fourth-order symmetric with $A_{1,1} = 0$. This kernel k^* may take a negative value, and hence, $\widetilde{F}_n(x)$ is not monotone as a function of x. However, the main purpose is to test the null hypothesis H_0 and to construct the confidence interval; that means we do not need to worry about it. As mentioned above, $\widetilde{F}_n(x)$ loses monotonicity at most 3% of its points around the origin when $n = 10$. For the smoothed Wilcoxon's rank test, we need only assume that the kernel k is fourth-order symmetric. While it is theoretically possible to construct a polynomial-type kernel that satisfies $A_{1,1} = A_{1,3} = 0$, it is rather complicated to do so, and it takes a couple of pages to write out the full form. Thus, we will only consider the kernel k^* here. It may be possible to construct another type of kernel that satisfies $A_{1,1} = 0$ and $A_{1,3} = 0$. We postpone this endeavor to a future work.

If the equation

$$V_0(\widetilde{S}) = \frac{n}{4} + o(1)$$

holds, we can use the Edgeworth expansion of $\widetilde{S}$ for testing H_0 and constructing a confidence interval without making any estimators. We can get an approximation of the α-quantile ($P_0(\widetilde{S} \leq \widetilde{s}_\alpha) = \alpha + o(n^{-1})$), i.e.,

$$\widetilde{s}_\alpha = \frac{n}{2} + \frac{\sqrt{n}}{2} z_\alpha + \frac{1}{48\sqrt{n}}(z_\alpha^3 - 3z_\alpha). \tag{5.16}$$

For the significance level $0 < \alpha < 1$, if the observed value $\widetilde{s}$ satisfies $\widetilde{s} \geq \widetilde{s}_{1-\alpha}$, we reject the null hypothesis H_0. Since the distribution function of $X_i - \theta$ is $F(x)$, we can construct the confidence interval of θ by using Eq. (5.16). For the observed value

$\mathbf{x} = (x_1, \ldots, x_n)$, let us define

$$\widetilde{s}(\theta|\mathbf{x}) = n - \sum_{i=1}^{n} K\left(\frac{\theta - x_i}{h_n}\right),$$

$$\widehat{\theta}_U = \arg\min_{\theta} \left\{\widetilde{s}(\theta|\mathbf{x}) \leq \widetilde{s}_{\alpha/2}\right\}$$

and

$$\widehat{\theta}_L = \arg\max_{\theta} \left\{\widetilde{s}_{1-\alpha/2} \leq \widetilde{s}(\theta|\mathbf{x})\right\},$$

where $0 < \alpha < 1$. The $1 - \alpha$ confidence interval is given by $\widehat{\theta}_L \leq \theta \leq \widehat{\theta}_U$.

Similarly, if the observed value $\widetilde{w}$ satisfies $\widetilde{w} \geq \widetilde{w}_{1-\alpha}$, we reject the null hypothesis H_0, where

$$\widetilde{w}_\alpha = \frac{n(n+1)}{4} + \frac{n\sqrt{2n+3}}{2\sqrt{6}}\left\{z_\alpha + \frac{1}{n}\left(\frac{7}{20}z_\alpha^3 - \frac{21}{20}z_\alpha\right)\right\}. \tag{5.17}$$

Using $\widetilde{w}_\alpha$ in (5.17), we can construct the confidence interval of θ. For the observed value $\mathbf{x} = (x_1, \ldots, x_n)$, let us define

$$\widetilde{w}(\theta|\mathbf{x}) = \frac{n(n+1)}{2} - \sum_{1 \leq i \leq j \leq n} K\left(\frac{2\theta - x_i - x_j}{2h_n}\right),$$

$$\widehat{\theta}_U^* = \arg\min_{\theta} \left\{\widetilde{w}(\theta|\mathbf{x}) \leq \widetilde{w}_{\alpha/2}\right\}$$

and

$$\widehat{\theta}_L^* = \arg\max_{\theta} \left\{\widetilde{w}_{1-\alpha/2} \leq \widetilde{w}(\theta|\mathbf{x})\right\}.$$

Thus, we have the $1 - \alpha$ confidence interval $\widehat{\theta}_L^* \leq \theta \leq \widehat{\theta}_U^*$.

Table 5.9 compares the simple normal approximation and Edgeworth expansion using the kernel k^* and the bandwidth $h_n = n^{-1/3}(\log n)^{-1}$. Since we do not know the exact distributions of the smoothed sign test $\widetilde{S}$, we estimated the values $P\left(\frac{\widetilde{S} - E_0(\widetilde{S})}{\sqrt{V_0(\widetilde{S})}} \geq z_{1-\alpha}\right)$ from 100,000 replications of the data and denote them as "True" in the table. "Edge." and "Nor." denote the Edgeworth and simple normal approximations, respectively. The underlying distributions are normal, logistic, and double exponential ones. The double exponential distribution is not differentiable at the origin (zero), but as mentioned before, we don't have to worry about that.

Remark 5.8 If we use a symmetric fourth-order kernel, which satisfies $A_{1,1} = 0$, the $n^{-1/2}$ term of the Edgeworth expansion is zero, and hence, the simple normal approximation means that the residual term is already $o(n^{-1/2})$. Comparing the $n^{-1/2}$ terms, we can see that the effect of the n^{-1} term is small; thus, the Edgeworth expansion with the $o(n^{-1})$ residual term is comparable to the simple normal approximation when the

Table 5.9 Comparison of normal approximation and Edgeworth expansion with the kernel $A_{1,1} = 0$ $(k^*, \ h_n = n^{-1/3}(\log n)^{-1})$

$\tilde{s}$	$n = 30$	$A_{1,1} = 0$		$\tilde{s}$	$n = 30$	$A_{1,1} = 0$	
$z_{0.99}$	True	Edge.	Nor.	$z_{0.95}$	True	Edge.	Nor.
N(0,1)	0.00842	0.01021	0.01	N(0,1)	0.05013	0.04993	0.05
Logis.	0.00937	0.01021	0.01	Logis.	0.0491	0.04993	0.05
D.Exp.	0.00908	0.01021	0.01	D.Exp.	0.04903	0.04993	0.05
$\tilde{s}$	$n = 100$	$A_{1,1} = 0$		$\tilde{s}$	$n = 100$	$A_{1,1} = 0$	
$z_{0.99}$	True	Edge.	normal	$z_{0.95}$	True	Edge.	Nor.
N(0,1)	0.00962	0.01006	0.01	N(0,1)	0.04903	0.04998	0.05
Logis.	0.00954	0.01006	0.01	Logis.	0.04892	0.04998	0.05
D.Exp.	0.0099	0.01006	0.01	D.Exp.	0.04937	0.04998	0.05

Table 5.10 Coverage probabilities of S, $\tilde{S}$, W and $\tilde{W}$ $(k^*, \ h_n = n^{-1/3}(\log n)^{-1})$

$n = 10, 90\%$	$s(\theta|\mathbf{x})$	$\tilde{s}(\theta|\mathbf{x})$	$w(\theta|\mathbf{x})$	$\tilde{w}(\theta|\mathbf{x})$
N(0,1)	0.9760	0.8910	0.9145	0.8910
Logis.	0.9788	0.8990	0.9158	0.8970
D.Exp	0.9806	0.8938	0.9189	0.9024
$n = 20, 90\%$	$s(\theta|\mathbf{x})$	$\tilde{s}(\theta|\mathbf{x})$	$w(\theta|\mathbf{x})$	$\tilde{w}(\theta|\mathbf{x})$
N(0,1)	0.9610	0.8921	0.9035	0.8982
Logis.	0.9610	0.8922	0.9026	0.8971
D.Exp	0.9589	0.8966	0.9055	0.8984

sample size n is small. When the sample size is large, the Edgeworth approximation is better.

Finally, we simulated the coverage probabilities based on S, $\tilde{S}$, W, and $\tilde{W}$, by making 100,000 repetitions. We used the intervals $\hat{\theta}_L \leq \theta \leq \hat{\theta}_U$ and $\hat{\theta}_L^* \leq \theta \leq \hat{\theta}_U^*$, where the kernel k^* and $h_n = n^{-1/3}(\log n)^{-1}$. For S and W, we constructed conservative confidence intervals whose coverage probabilities are equal to or greater than $1 - \alpha$ when the sample size is small. Table 5.10 shows that the coverage probabilities of the smoothed statistics are less conservative.

Remark 5.9 If the sample size n is large enough, the higher order approximation works well. In that case, we recommend the Edgeworth expansion with the pair $(k^*, \ h_n = n^{-1/3}(\log n)^{-1})$. If the sample size is moderate, the normal approximation based on the pair $(k_{e,4}, \ h_n = n^{-1/3})$ works well. In that case, we recommend the fourth-order kernel and bandwidth $h^{-1/3}$.

References

1. Evans DL, Drew JH, Leemis LM (2017) The distribution of the Kolmogorov-Smirnov, Cramer-von Mises, and Anderson-Darling test statistics for exponential populations with estimated parameters. Comput Probab Appl 247:165–190
2. Finner H, Gontscharuk V (2018) Two-sample Kolmogorov-Smirnov-type tests revisited: old and new tests in terms of local levels. Ann Stat 46(6A):3014–3037
3. Baringhaus L, Henze N (2016) Cramér-von Mises distance: probabilistic interpretation, confidence intervals, and neighbourhood-of-model validation. J Nonparametric Stat 29:167–188
4. Curry J, Dang X, Sang H (2019) A rank-based Cramér-von-Mises-type test for two samples. Braz J Probab Stat 33(3):425–454
5. Chen H, Döring M, Jensen U (2018) Test for model selection using Cramér-von Mises distance in a fixed design regression setting. Adv Stat Anal 102:505–535
6. Omelka M, Gijbels I, Veraverbeke N (2009) Improved kernel estimation of copulas: weak convergence and goodness-of-fit testing. Ann Stat 37:3023–3058
7. Brown B, Hall P, Young G (2001) The smoothed median and the bootstrap. Biometrika 88(2):519–534
8. Maesono Y, Moriyama T, Lu M (2016) Smoothed nonparametric tests and their properties. ArXiv:1610.02145
9. García-Soidán PH, González-Manteiga W, Prada-Sánchez J (1997) Edgeworth expansions for nonparametric distribution estimation with applications. J Stat Plan Inference 65(2):213–231
10. Huang Z, Maesono Y (2014) Edgeworth expansion for kernel estimators of a distribution function. Bull Inform Cybern 46:1–10
11. Bickel P, Götze F., van Zwet W (1986) The edgeworth expansion for U-statistics of degree two. Ann. Stat. 14:1463–1484
12. Loeve, M.: Probability Theory. Van Nostrand-Reinhold, New Jersey (1963)

Printed in the United States
by Baker & Taylor Publisher Services